CHANGING CHARACTER OF

HYBRID WAR

The Threat to India

CHANGING CHARACTER OF HYBRID WAR

The Threat to India

Col. Shubhankar Basu, SM, VSM

Vivekananda International Foundation
New Delhi

PENTAGON PRESS LLP

Changing Character of Hybrid War: The Threat to India

By Col. Shubhankar Basu, SM, VSM

ISBN 978-81-944659-8-0

First Published in 2020

Disclaimer: The views expressed and suggestions made in the book are solely of the author in his personal capacity and do not have any official endorsement. Attributability of the contents lies purely with author. The publisher does not take any responsibility for the same in any manner whatsoever.

Published by
PENTAGON PRESS LLP
206, Peacock Lane, Shahpur Jat,
New Delhi-110049
Phones: 011-64706243, 26491568
Telefax: 011-26490600
email: rajan@pentagonpress.in
website: www.pentagonpress.in

Printed at Aegean Offset Printers, Greater Noida, U.P.

Contents

Foreword

Warfare has never been purely military. It has always been hybrid, involving the domains of statecraft, diplomacy, information, economics, law, social engineering and many others. What fighting doctrines, weapons, technology, proxy votes, irregular warfare, propaganda, economic and legal manoeuvres have been an integral part of hybrid warfare.

Since its independence, India has fought four military conflicts, apart from having been a victim of terrorist attacks, proxy wars, propaganda wars, cyber-attacks, fake news, et cetera. India has been constantly fighting in hybrid warfare mode. The Indian Army doctrine of 2018 recognises that the future conflicts will be "hybrid in nature, a blend of conventional and unconventional, with the focus increasingly shifting to multi-domain warfare varying from non-contact to contact warfare." India has for long been facing a collusive threat from China and Pakistan. This will deepen further in future.

The information dimension of hybrid warfare needs special attention. The breathtaking advances in information and application technologies, the emergence of social media platforms which facilitate instantaneous outreach to millions of intended targets, encryption technologies which provide anonymity on the social media platforms, cyber attacks, the emergence of new technologies like artificial intelligence which can be used for making autonomous weapon systems, the advances in sensors technologies have made hybrid warfare even more deadly. More lives are lost through terrorism, activation of a country's fault lines, instigation of riots, disruption of livelihoods. The global pandemic Covid 19 has caused hundreds and thousands of deaths in a matter of a few weeks and months. A fierce blame game has erupted between the US and China as to who was responsible for the spreading of the coronavirus globally. The pandemic has once again focused attention on the bioterrorism and the use of viruses and bacteria in warfare

Most militaries around the world are amending their warfighting doctrines to fight in the grey zone conflicts in which perception and reality are difficult to distinguish, the targets are invisible, and the attacks on self are silent. The kinetic nature of warfare is complemented by the information warfare which is fought relentlessly 24x7. The biggest challenge before the militaries is to prepare themselves to fight the multi-domain hybrid conflicts, also known as grey zone conflicts, and survive them. Fighting such conflicts requires a whole of society approach, the military is just one of the many domains of the conflict.

Col Shubhankar Basu of the Indian Army spent a year at the Vivekananda International Foundation researching hybrid warfare. He has produced a book titled *CHANGING CHARACTER OF HYBRID WAR: The Threat to India.* The Vivekananda International foundation is happy to publish Colonel Basu's work.

The volume goes into the history of hybrid warfare, examines the perspectives of leading countries and looks at the evolution of the concept through a variety of case studies including the Israel Hezbollah conflict, cyber attacks on Estonia, Georgia-Russia war, the rise of Islamic state et cetera. Two chapters look at Pakistan's and China as concepts and practices in hybrid warfare. The final chapter is about understanding the collusive threat from China and Pakistan to India and how India is coping with this.

The book also provides useful recommendations and suggestions. The author recommends a holistic all of the government approach for effectively dealing with the collusive hybrid warfare threat to India. Given the multi-domain nature of hybrid warfare and the involvement of myriads of government departments and agencies, the author suggests the establishment of a nodal agency by an act of Parliament to deal with hybrid warfare. The present structure of the National Security Council is not sufficient as its recommendations are not binding. It does not even meet regularly at even at the time of crisis. Col Basu also suggests the integration of cyber, space and electronic warfare assets in due course, the creation of a National Centre for Counterterrorism, early operationalisation of National Grid agency and police reforms which have been delayed for too long. Some of the suggestions like the creation of the Chief of Defence Staff, have been implemented. Some, like the preparation of an authorised national security strategy, have yet to be implemented.

National security threats are ever-evolving. They can be no one prescription for all times. The government should have the political will, the resources and the requisite institutions and mechanisms to deal with the changing nature of hybrid warfare. That is the key message of Col Basu's book.

New Delhi
7 May 2020

Dr Arvind Gupta
Director, Vivekananda International Foundation

Acknowledgements

I would like to express my sincere gratitude to the Vivekananda International Foundation for providing an opportunity to undertake research for this book. I am grateful to the VIF for allowing me to be part of the enriching discussions during the meetings of the Pakistan Study Group, the China Study Group and the numerous round table discussions on the contemporary international security situation.

This book has been possible especially due to the guidance of Dr. Arvind Gupta, Director VIF, who not only reposed confidence in me to undertake this project but has also been a constant source of inspiration.

I am grateful to Lt. Gen. Ravi Sawhney, PVSM, AVSM who has been my guide and mentor during the study. Without his guidance and constant encouragement, this book would not have been possible. I take this opportunity to thank Lt. Gen. Gautam Banerjee, PVSM, AVSM, YSM; Maj. Gen. P.K. Mallick VSM; Brig. Anand; Prof. Sujit Datta; Dr. Sreeradha Dutta and Dr. Harinder Sekhon for their support and guidance throughout my research. A special thanks goes to Commodore Somen Banerjee, who was more than a senior colleague. He played the role of mentor, sounding board, critic, and most important, was like the seniors from the Academy days who would push their juniors hard to get the best out of them.

My gratitude and compliments to the young scholars who helped me with resource material and reviewing and helping me refining my work, with special mention of Prateek, Dr. Teshu, Prerna, Ayush, Kapil and Anurag. I wish to thank Ms. Jyothi J. Nair (GIS consultant) for helping me with the maps and Shri Krishan Redhu who provided the IT support.

My special gratitude to Lt. Gen. Ata Hassnain and Brig. R.K. Bhonsle SM who reviewed my study report and made valuable recommendations.

Last, but not the least, the book would not have taken its present shape without the support and suggestions of Mrs. Anuttama Ganguly.

Col. Shubhankar Basu

List of Acronyms

ADIZ	Air Defence Identification Zone
AGPL	Actual Ground Position Line
AGPL	Actual Ground Line of Patrolling
AI	Artificial Intelligence
ANA	Afghan National Army
ANDSF	Afghan National Defence Security Forces
AQI-	Al-Qaeda in Iraq
ASAT	Anti-Satellite Strike Capability
ASEAN	Association of South East Asian Nations
ATM	Automated Teller Machine
BCDA	Border Defence Cooperation Agreement
BIMSTEC	Bay of Bengal Initiative for Multi-sectorial Technical and Economic Cooperation
BRI	Belt and Road Initiatives
BSNL	Bharat Sanchar Nigam Limited
CBI	Central Bureau of Investigation
CBMs	Confidence Building Measures
CCS	Cabinet Committee on Security
CDS	Chief of Defence Staff
CERT	Computer Emergency Response Team
CFR	Council on Foreign Relations
CLAWS	Centre for Land and Warfare Studies
CMC	Central Military Commission
COAS	Chief of the Army Staff
COG	Centre of Gravity
CPC	Communist Party of China
CPEC	China Pakistan Economic Corridor
CPI (M)	Communist Party of India (Marxist)

CRBC	China Road and Bridge Corporation
CRPF	Central Reserve Police Force
CRS	Congressional Research Service
CSTO	Collective Security Treaty Organization
CUW	Counter Unconventional Warfare
Dodos	Distributed Denial of Service
DG ISPR	Directorate General of Inter Services Public Relations
DGP	Directorate General of Police
DNS	Domain Name System
DDoS	Denial of Service
EBR	East Bengal Rifles
EEZ	Exclusive Economic Zone
ETH	Ethereum
EU	European Union
EW	Electronic Warfare
FAT	Falah-e-Aam Trust
FATA	Federally Administered Tribal Area
FATF	Financial Action Task Force
FICN	Fake Indian Currency Notes
FIR	First Information Report
GoM	Group of Ministers
GPD	General Political Department
GSD	General Staff Department
GWOT	Global War on Terror
HM	Hizbul Mujahideen
HQ	Head Quarter
HUJI	Harkat-ul-Jihad al-Islami
IB	Intelligence Bureau
IDF	Israeli Defence Forces
INCSR	International Narcotics Control Strategy Report
IOR	Indian Ocean Region
IoT	Internet of Things
IPC	Indian Penal Code
ISI	Inter-Services Intelligence
ISIS	Islamic State in Iraq and Syria
ISPs	Internet Service Providers
IT	Information Technology
IW	Information Warfare
JeM	Jaish-e-Mohammad

JKART	Jammu Kashmir Affectees Relief Trust
JKLF	Jammu and Kashmir Liberation Front
JMB	Jamaat-ul-Mujahideen Bangladesh
JUI	Jamaat-e-Ulema Islam
KKH	Karakoram Highway
LAC	Line of Actual Control
LC	Line of Control
LeT	Lashkar-e-Taiba
LOC	Line of Control
LTC	Lite coin
LTTE	Liberation Tigers of Tamil Elam
LWE	Left Wing Extremism
MAC	Military Area Commands
MCDC	Multinational Capability Development Campaign
MEA	Ministry of External Affairs
MHA	Ministry of Home Affairs
MOD	Ministry of Defence
MPECI	Military Political Economic Civil and Information
MQM	Mohajir Qaumi Movement
MR	Military Regions
MRBM	Medium Range Ballistic Missile
MTCR	Missile Technology Control Regime
MTNL	Mahanagar Telecom Nigam Limited
NATO	North Atlantic Treaty Organization
NCCC	National Cyber Coordination Centre
NCSC	National Cyber Security Coordinator
NDPS	Narcotic Drugs and Psychotropic Substances
NDS	National Defence Strategy
NIA	National Investigation Agency
NMO	National Military Objectives
NSAB	National Security Advisory Board
NSC	National Security Council
NSCN-IM	National Socialist Council of Nagaland–Isak Muivah
NSD	Networks System Department
NWFP	North West Frontier Province
NWNP	No War and No Peace
OGWs	Overground Workers
OHCHR	Office of the United Nations High Commissioner for Human Rights

OIC	Organization of Islamic Countries
PCA	Permanent Court of Arbitration
PLA	Peoples Liberation Army
PMESII	Political Military Economic Social Infrastructure and Information
POK	Pakistan Occupied Kashmir
PQ	Parliament Questions
PRC	Peoples Republic of China
PTM	Pashtun Tahaffuz Movement
QDR	Quadrennial Defence Review
RMA	Revolution in Military Affairs
SAF	Singapore Armed Forces
SCDF	Singapore Civil Defence Force
SCO	Shanghai Cooperation Organization
SCS	South China Sea
SEZs	Special Economic Zones
SOF	Special Operations Forces
SPG	Strategic Policy Group
SQL	Structured Query Language
SRBM	Short Range Ballistic Missile
SSD	Space Systems Department
SSF	Strategic Support Force
TAIEX	Taiwan Stock Exchange
THAAD	Terminal High Altitude Area Defence
UAE	United Arab Emirate
UK	United Kingdom
ULFA	The United Liberation Front of Assam
UN	United Nations
UNCLOS	United Nations Convention on Laws of Sea
UNLF	United Liberation Front of Assam
US	United States
US CENTCOM	United States Central Command
USA	United States of America
USSR	United Soviet Socialist Republics
UWSA	United Wa State Army
WWII	World War II
ZEC	Zcash

Introduction

> *"Both the external and internal environment are changing at an incredibly fast pace, with developments in nuclear weapons and missiles, increasing cross-border terrorism, the emergence of non-state actors, the growth of Islamic fundamentalism, the narcotics-arms nexus, illegal migration and left-wing extremism, gravely impacting upon the security of the country."*
>
> **—The Kargil Review Committee (KRC) Report 2001**

The Kargil Review Committee (KRC) in its report in 2001 had visualized that the internal and external environment are changing at an incredible fast pace. The trajectory of the character of war has evolved even faster with technology being the prime driver. The emergence of fourth and fifth generation warfare and direct conflict between the non-state actors and states has blurred the distinction between traditional war and peace. The Indian Army 2018 doctrines defines the primary security challenge to India as "*The greatest danger to our Nation continues to be external aggression with collusive threat from adversaries as well as internal instability due to state/non-state sponsored proxy war, thus imposing multi front challenges in various dimensions, ranging from warfare in the non-contact domain, to an all-out full spectrum war.*" It further amplifies the changing character of war, "*the future conflicts will be characterised by operating in a zone of ambiguity where nations are neither at peace nor at war—a 'Grey Zone' which makes our task more complex. Wars will be hybrid in nature, a blend of conventional and unconventional, with the focus increasingly shifting to multi domain warfare varying from non-contact to contact warfare.*"The hostile actions of our adversaries can be summarised as hybrid warfare as these actions span the complete spectrum of conflict. The sponsoring of terrorism and proxy war in Jammu and Kashmir, launching of cyber-attacks, pushing in of narcotics and counterfeit currency, lie at the lower end of the spectrum of conflict, the combat operations on Line of Control and threat of irrational use of nuclear weapons are at the other end of the conflict.

The collusive threat from China and Pakistan would imply that both countries are acting together covertly to achieve a fraudulent, illegal or deceitful goal against India or to threaten its territorial integrity. The China-Pakistan collusion began in 1963, when Pakistan and China signed the boundary agreement delimiting the boundary between Xinxiang and Northern Areas of PoK. While signing the agreement, Pakistan ceded Shaksgam valley to China. The Pakistan-China military collusion can manifest either in the form of providing support for war-less physical military intervention or by mobilising PLA and undertaking a military intervention against India. Former COAS Gen V. P. Malik had said that the transition from one form of collusion to the other could take place seamlessly.[1] China has provided military material support in terms of high technology weapon systems and force multipliers. The ongoing CPEC project has further enhanced the cooperation between the two nations. In a conflict scenario between Pakistan and India, apart from providing support in information warfare and real-time tactical intelligence, a mere positioning or mobilisation of PLA troops by China in Tibet can adversely affect the force equilibrium for India. China may not be overtly supporting Pakistan in its proxy war against India in the present context, but would continue assisting Pakistan diplomatically, militarily and economically in all its misadventures. China is unlikely to make any positive move to resolve the boundary issue with India; rather it will keep the LAC active by orchestrating face-offs, which will keep the Indian Army engaged along the LAC. The collusion between China and Pakistan in the domain of hybrid warfare is keeping India engaged in security-related issues both internally and on the borders.

A large number of scholars have written about the new forms of warfare which are just below the threshold level of conventional or all-out war and marginally above the level of conflict. The term 'hybrid warfare', meaning simultaneous employment of two or more methods to achieve the desired result has generated research and discussion as well as criticism amongst scholars, practitioners and experts. The term may have come into vogue in the last few years but there is no novelty in the concept and the phenomenon. Ancient strategic thinkers and philosophers such as Kautilya and Sun Tzu have propagated and executed hybrid warfare. Post-World War II there has been a substantial change in the character of conflict. This may be attributed to two factors—first was the advent of nuclear power (technological cause) and the cold war between the USA and USSR (political). Martin Van Creveld in his book, '*The Transformation of War*' has argued that nuclear armed states have not gone for a war since 1945;[2] even the Allies who have 'protection' have been immune to conventional war. The most important wars since 1945 have been between non-nuclear powers. The USA-led, Gulf wars I & II and Global War on Terrorism (GWOT-2001 onwards) witnessed the quantum change in technology and tactics being used in conflict. Post the Cold War, use of irregulars and sub conventional means like psychological, religious and information warfare simultaneously and in a synchronised manner brought in a transformation in the character of war. The wars in Estonia (2007), Georgia (2008) and Crimea (2015) indicate the effectiveness of irregular, information and cyber warfare by achieving decisive results in a rather quick-time frame. The Lebanon war of 2006 and the campaign of the ISIS in Iraq and Syria since 2014 indicate a blending of lethality and intensity of state conflict with the fanatical and proactive fervour of irregular warfare. China has successfully used a combination of economic appeasement, legal and psychological warfare along with coercion by military power to achieve its national aims in the South and East China Seas. The offensive and military use of drones in the Iran-US standoff, the strike on Saudi oil facilities and recent dropping of weapons and other war-like stores in areas close to the Pakistan border in Punjab has added a new aerial dimension to the evolving character of warfare. The increasing use of cyber, information, religious, irregular, legal, psychological and economic means in a near simultaneous timeframe to achieve national aims indicates the maturing of hybrid warfare strategy.

This book titled 'Changing Character of Hybrid War: The Threat to India' examines the emergence of hybrid warfare and analyses the collusive hybrid threat to India. The book comprises five chapters. The first titled 'Hybrid Warfare: A Plural Perspective', examines the emergence of hybrid warfare through the writings of Kautilya, Sun Tzu, Clausewitz and Martin Van Creveld. The literature and doctrine of major powers have been analysed to study respective perspectives on the emergence of hybrid warfare. This chapter argues that the transformation in the character of war has manifested in the emergence of hybrid wars. Chapter two titled 'Transformation of Hybrid Warfare' explores the transformation of hybrid warfare through case studies of conflicts since post-cold war. The emerging character and future trends in hybrid war including the impact of disruptive technologies have also been analysed. Chapter three titled 'Pakistan an Ardent Practitioner of Hybrid Warfare' analyses Pakistan's interference in the internal affairs of Afghanistan and the turbulent relations it has with India since its creation.

This chapter argues that the strategic culture of the Pakistan Army is the genesis of the turbulent relationship that Pakistan has with India and Afghanistan. Chapter four titled 'Chinese Characters of Hybrid Warfare' delves into Chinese white papers on defence to study the evolution of Chinese military strategy and transformation of the PLA and its convergence with hybrid warfare. The use of 'three warfare strategy and non-kinetic means' by the Chinese to achieve its national aims has been examined through the case studies of 'Chinese coercion in the South and East China Seas' and use of 'Economic Warfare.' Chapter five titled 'Collusive Hybrid Threat to India' discusses the current and emerging faultiness which can be exploited by adversaries and delves into contours of collusive hybrid threat to India. This chapter argues that there is need to build deterrence against hybrid threat and proposes measures which may be considered to counter the collusive hybrid threat in future.

> *'The constant threat of terrorism and the low intensity proxy war has become the disturbing features of national life. This constitutes the new face of war.'*
>
> **—Dr. A. P. J. Abdul Kalam**

NOTES

1. General VP Malik, PVSM, AVSM (Retd.) former COAS, during the 30th USI National Security Lecture in 2014 available at https://indianarmy.nic.in/WriteReadData/Documents/national%20Security%20Lecture%202014.pdfaccessed on Jul 10 2019.
2. Less Russia-China border skirmish of 1969. The incident cannot be categorized as a war as none of the parties declared it as such.

1

Hybrid Warfare: A Plural Perspective

"Future conflicts will be characterised by operating in a zone of ambiguity where nations are neither at peace nor at war—a 'Grey Zone' which makes our task more complex. Wars will be hybrid in nature, a blend of conventional and unconventional, with the focus increasingly shifting to multi domain warfare varying from non-contact to contact warfare."[1]

—Indian Army Land Warfare Doctrine - 2018

ABSTRACT

Is hybrid warfare a new concept or just semantics? What is new in hybrid warfare? Is there a need to study and prepare differently? These are some of the questions that have been deliberated in this chapter through the writings of Kautilya, Sun Tzu, Clausewitz and Martin Van Creveld. The literature and doctrine of major Powers have been analysed to bring out the respective perspective on the emergence of hybrid warfare. This chapter argues that the transformation in the character of war has manifested in hybrid wars. The major Powers may view the hybrid war differently but converge on to the proposition that there is need to train, equip and prepare differently for hybrid warfare.

The proposition that wars or the organised violence of 21st century is different from the wars of the 20th century has generated widespread discussions and has been debated in not only in the military but also in the scholarly and policy literature.[2] Various terms have been used to conceptualise the contemporary conflict-wars among the people, like wars of the third kind, hybrid wars, postmodern wars as well as 'new wars'.[3] A scan of conflicts in the current and last century reveals that the scope, dimensions or the means of war fighting have been changing too fast to establish or define the changing trend. Conventional war, nuclear war, sub-conventional war, asymmetric war, total war, nonlinear war, irregular war, compound war, non-conventional war, non-contact warfare, and hybrid war, are a few of the various ways in which scholars or military minds have classified the different ways of conflicts or wars that are being fought. The only common thread in the abovementioned categories of the conflict is possibly the violence. The space between 'War' and 'Peace' is occupied by various shades of violence. Violence has been defined by the World Health Organisation as the intentional use of physical force or power, threatened or actual, against oneself, another person or against a group or community, which either results in or has a high likelihood of resulting in injury, death, psychological harm, mal-development or deprivation.[4] Violence falls into essentially two forms: random violence, which includes actions carried out with sanction (as in war) or unsanctioned violent groups (that is, inter-societal violence) and terrorism.[5] Post-cold war, the violence is no more restricted to the two forms as discussed above. Now transnational organisations like Al Qaeda and ISIS are challenging the nation states using all means of violence, akin to war.

> *Hybrid wars are 'full spectrum wars with both physical and conceptual dimensions: the former, a struggle against an armed enemy and the latter, a wider struggle for, control and support of the combat zone's indigenous population, the support of the home fronts of the intervening nations, and the support of the international community.'*
>
> **—J. J. McCuen**

A large number of scholars have written about the new forms of warfare which are just below the threshold level of conventional or all-out war and marginally above the level of conflict. Frank G. Hoffman, in his seminal book, *Conflict in 21st Century: Rise of Hybrid Wars*, discussed Hybrid Warfare in detail. The relevance and understanding of hybrid warfare has become enriched consequent to the criticism it received on being quoted as a new form of warfare.[6] The criticism is primarily based on the meaning of hybrid, which means simultaneous employment of two or more methods to achieve the desired result; hence there is no novelty in the concept. Wars have always been fought using two or more means. This criticism also may be arising out of the fact that different commentators have defined hybrid warfare in different ways.[7] As per Hoffman, competitors with different approaches (conventional, irregular, terrorist) will employ all forms of war and tactics, perhaps simultaneously. According to J.J. McCuen, hybrid wars are 'full spectrum wars with both physical and conceptual dimensions: the former, a struggle against an armed enemy and the latter, a wider struggle for control and support of the combat zone's indigenous population, the support of the home fronts of the intervening nations and the support of the international community.'[8] Is hybrid warfare just semantics or merits to be studied as a different form of war? The issue needs to be analysed against the context of evolving the aims of war, the emerging actors of war, the transformations in method of fighting the war and new routes of financing. This chapter will examine the concept and definition of war/hybrid war as given out by various philosophers and scholars, the doctrines and official notes of major armies of the world to establish the need to study hybrid war as a different form of war. A time correlation matrix in graphical form is given below (Figure 1.1) to understand the environment that prevailed when these philosophers and scholars wrote about 'War'.

Figure 1.1: Chart showing the period during which the various scholars wrote about war

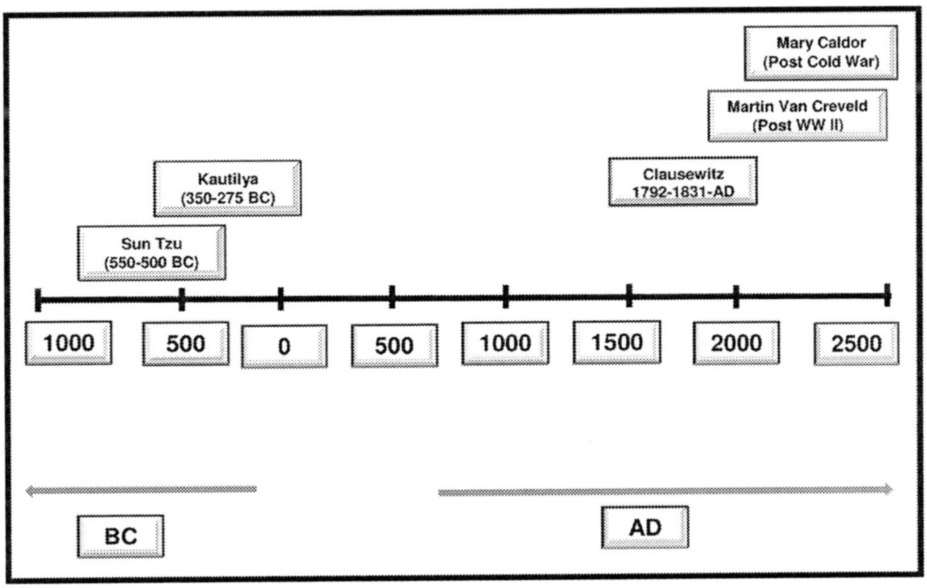

Historical Perspective. Martin Van Creveld in his book '*More on War*' says that the two great military historians who have been read the most and widely understood are Chinese sage Sun Tzu (544-496 BC) and the Prussian soldier-philosopher, Carl Von Clausewitz (1780-831).[9] He goes on to say that, most of the other military historians were considered too philosophical or outdated.

Surprisingly the great Indian philosopher Kautilya does not find mention in the book of Martin Van Creveld when he is writing in the context of military theory and history. Clausewitz in '*On War*' says "... *War is the continuation of the politics,*"[10] and he defined war as "... *War is thus an act of force to compel our enemy to do our will.*"[11] Thus, '*imposing of the will*' becomes an objective of war and 'use of force' is the means to achieve the same. However, the nature and quantum of force to achieve the objective remains a subject of enquiry which will be discussed later in the chapter. Hence, if the objective of war is to achieve the national aim, then it would be prudent to analyse what Sun Tzu and Kautilya have written. According to Sun Tzu '*...to fight and conquer in all your battles is not the supreme excellence; excellence consists of breaking enemy's resistance without fighting.*'[12] The master strategist laid emphasis on achieving the aim of breaking the enemy's resistance without fighting.

Kautilya in his master treatise *Arthashastra* refers to four types of wars:[13]

- **Mantra Yuddha or War by Counsel.** This is the exercise of diplomacy to win wars. It has to be utilized by the King when he is not in the favourable position to launch a war or it is not beneficial for him to go for battle.
- **Prakasa Yuddha or Open War.** This is a form of conventional war which follows all the rules of fighting a battle. Kautilya advises the King that "when he is superior in troops and when he is on the land suitable to himself, he should engage in an open fight..."
- **Kuta Yuddha or Concealed War.** This is irregular warfare by use of covert means like treachery in the enemy camp or laying ambushes in the rear.
- **Guda Yuddha/Tusnim Yuddha or Clandestine/Silent War.** This type of war is waged by covert means to achieve the objective without actually waging the battle, like assassinating the enemy. This would be generally done in peace time. The King may not only use his agents but also allies, vassal states, friends and enemy's friends.

Of the four types of warfare suggested by Kautilya, in three, employing physical force, using violence or fighting of war is not professed. Only 'Prakasa Yuddha or Open War' is a full-fledged war. The means to achieve the will of the state is achieved by means which may be other than physical war or violence. However all four types are still called war. This highlights the point that in ancient India the definition of war really did not connect with violence. The teachings of Sun Tzu have a similar resonance when he talks of defeating the enemy without fighting.

None of the two great historians mentioned by Martin Van clearly bring out the cause of war. While, Sun Tzu in his legendary work, "***The Art of War***" says, "*War is of vital importance to state*"[14] and brings out the importance of military studies by saying that "*...is subject of enquiry which on no account can be neglected.*" The legend then brings out the factors governing the war followed by laws governing them, conduct and preparation.[15] He does not specifically establish the cause or reasons that push states to war or the reasons for which the king or emperor seek war. It is important to discuss the cause or reasons for which states go to war because it is important in understanding the taxonomy of the changing character of war.

As per Clausewitz "*... War is thus an act to force to compel our enemy to do our will*".[16] It is further amplified as nothing but a duel on a larger scale, wherein each fighter uses force to compel the opponent to follow his will. Maybe a thousand or more duels be necessary to qualify for a war, but lesser numbers, may be called a conflict.[17] However, later in his book Clausewitz specifies violence as a major characteristic of war but then it can be argued considering the context under which the book was written.[18] The initial books (01 to 06) of '*On War*' were influenced by the French wars (1792-1815: The Napoleonic era) which are influenced by victory in the battle. While reviewing his books 1 & 2, he was in the process of reviewing the definition of victory and mentioned that the connotation of the definition of victory has changed with time and it is partial or total destruction of enemy's forces or till he surrenders or to make him compel follow the victor's will.[19] Moreover, he could not complete his work and had got some more time there may have been revisions in his earlier books too. Hence, to restrict the definition of war with the existence of violence needs to be seen through the prism of the conditions when Clausewitz wrote '*On War*'. Napoleon's celebrated victory at Ulm,[20] where the opponent surrendered on being outflanked and outnumbered is called the perfect victory by many military historians and is an example of how through manoeuvring violence can be avoided.[21]

The last books of Clausewitz reflect his realisation that war could come in different shapes and forms[22] and the character of war is ever evolving. This is evident from the relevance of his concept of '*Trinity*' in the context of modern wars. Clausewitz identified the concept of the trinity of variables, the relationship between government, military and population (the violence-hatred-enmity was associated with the passion of people), which became the fundamental tool for the study of war for scholars in the 20th and 21st centuries.[23] The wars of the 21st century have blurred the distinction between the government, military and population.[24] Most of the wars of the 21st century are being fought between networks of state and non-state actors and it is near impossible to distinguish between non-state actors, combatants, hired combatants or mercenaries and civilians.[25] There are critics who have argued on the out datedness of Clausewitz's trinity as guerrilla forces, insurgents, transnational terrorist groups and warlords cannot be neatly categorised into government, military and population.[26]

In the construct of post-World War II (WW-II), the philosophy of war as given in the *Arthashastra* becomes more relevant. This Kautilyan convergence of diplomacy/means of political influence, i.e., (Four *Upayas*)—Sama (conciliation or diplomacy), *Dana* (economic gratification), *Bheda* (dissension or information operations) and *Danda* (use of force)[27] and other means of war to achieve the end state are aligned with the concepts of hybrid warfare (use of all elements of national power) as articulated by various scholars 2000 years later. Numerous examples such as China's Cheque Book Diplomacy, role of the USA in breaking up of the USSR, NATO air strikes on Yugoslavia, sanctions against North Korea or Iran are testimony of Kautilyan thought being more relevant today.

Clausewitz was not only influenced by Machiavelli's '*The Prince*', but was a great admirer of the seminal work where the prince uses war as one of the many tools for his political ends.[28] Martin Van Creveld brings out that neither Sun Tzu nor Clausewitz has established a relationship between economics and war.[29] While, Sun Tzu had emphasised "*...there is no instance of a country having benefited from a prolonged warfare.*"[30] However, there is no direct connect that economics could be the reason for commencement or end of a war. Since, WW II, more wars have an economic angle than territorial or ideological angles.[31] If the relevance of economics is the argument, then Kautilya in his treatise on statecraft and strategy, '*The Arthashastra*' uses '*Dana* meaning economic gratification' to achieve the end state.[32] The concept of state or an ideal state in *Arthashastra*, which Kautilya has used as a base to explain his theories can also be called the Kautilyan state, has trade and economics in the centre.[33] The guard posts are all around the centres of trade. Hence, the importance of economics to the role of state craft or king to guard the centres of economics finds its relevance since the time of Kautilya. Since the latter part of the 20th century, more wars have been fought for economic reasons than for ideology or geo-political reasons.[34]

The above analysis establishes the argument that warfare had a much larger definition and covered a wider spectrum during the time of Sun Tzu and Kautilya.

In the oriental or philosophy of the East, physical fighting or violence, came at a much later stage and was the last option. The Clausewitzian concept of use of violence to fight war or achieve the will of the nation was primarily influenced and dominated by the geopolitical and strategic environment existing in Europe at that time.

Emerging Contours of Warfare post World War II. The nature of war does not change, it is enduring. However, the character of war evolves or transforms not only due to technological advancement but also because of social, political and economic changes. Post World War II there has been a substantial change in the character of conflict. This may be attributed to two factors—first was the advent of nuclear power (technological cause) and the cold war between the USA and the USSR (political). Martin van Creveld in his book '*The Transformation of War*' argues that nuclear weapons were built to give the military and nation states unprecedented powerful tools for making and winning the war. However, nuclear weapons have actually put an end to the possibility of war between two nuclear armed nations.[35] He has argued that no nuclear armed states have gone for war since 1945[36] Even the Allies who have 'protection' have been immune to conventional war.[37] The most important wars since 1945 have been between non-nuclear powers, the Middle East (1948-49, 1952, 1967, 1973, 1982 and 1980-1988), between China and Taiwan (1954, 1958), the USA and Vietnam (1964), India and China (1962), India and Pakistan (1947, 1965 and 1971).[38] Numerous wars have been fought during the cold war period between the US-led NATO and the USSR-led Warsaw Pact countries. Invariably one of the superpowers was always behind the conflict. Hence, the major powers being nuclear did not opt for direct engagement, but indirect engagement continued. The indirect engagement could be from protecting human rights to regime change. Hence, the 'war of interest' which had evolved into a 'war of conscience' graduated to a 'war of intervention' by the early part of the 21st century.[39] Post World War II did not witness change only in terms of parties to war, but also in the intensity. Since 1945 the number of conventional wars at best would be just around 20, whereas sub-conventional wars (can be defined as conflicts waged by or against non-state actors, terrorism and insurgencies) in the same period would be well over 100.[40] Mary Caldor and Brig. Gurmeet in their respective essays have endorsed the issue discussed earlier that there has been a marked decrease in inter-state wars while the number of civil wars have been increasing since WWII.[41] The term 'subconventional war' evolved post World War II, as conventional war has been linked with employment of nuclear weapons. However war being a complex business, such simple classifications may not work, which has been proved by full scale wars between nation states since WW-II.

The Vietnam Wars (1946-54 & 1954-1974), Afghanistan (1979-89) and the USA-led Global War on Terrorism (GWOT-2001 onwards) witnessed a quantum change in technology and tactics being used in conflict. The use of irregulars and sub conventional means like psychological, religious and information warfare simultaneously in a synchronised manner brought in a transformation in the character of war. It was around the first decade of the 21st century that the term 'Hybrid Warfare' started appearing in military journals to explain the emerging trends of war. Frank Hoffman was one of the first scholars who characterised the term 'Hybrid Warfare.'[42] In his study, he concluded that hybrid wars blend the lethality and intensity of state conflict with the fanatical and protractive fervour of the irregular warfare.[43] The wars in both Georgia (2008) and Crimea (2015) indicate the effectiveness and maturing of hybrid warfare strategy by achieving decisive results in rather quick time.[44] The perspective of the USA, China, UK, Russia, and India have been analysed in subsequent paragraphs to arrive at an understanding of Hybrid Warfare.

> *The United States will fight 21st century war among "hybrid conditions" and emphasized the need to maintain focus on the mixed type of warfare and to make irregular war a core competency.*
>
> —**Gen Mattis**

U.S. Perspective: Hybrid Wars and Grey Zone Conflicts. The U.S. Department of Defence's *Quadrennial Defence Review* (QDR)-2006 was written after four long bloodied wars in Afghanistan which was irregular in nature. It was observed that the enemies are not conventional forces, but were widely displaced terrorist networks which use terror propaganda and indiscriminate violence to subjugate the Muslim world and exploit Islam to achieve radical political aims.[45] The QDR called for reorienting the capabilities of the U.S. military to address a wider range of challenges to include the non-traditional asymmetric threat to the country. Out of the four priority focus areas of QDR, defeating terrorist networks and providing security to the homeland in depth were the top two.[46] The QDR 2006 provided new direction transformation for the forces to focus more on terror-related activities, counter-terrorism, counter-insurgency and long duration irregular warfare. The National Defence Strategy (NDS)-2005 and later 2008, identified the strategic environment as a global struggle against violent extremist ideology that is transnational and threatens to overturn the international state system.[47] This was the time when scholars, academicians and military experts in the U.S.A. like Hoffman, John Arquilla and General Mattis (and many others) started speaking of 'Hybrid Warfare'. Gen Mattis said "*the United States will fight 21st century war among "hybrid conditions"*" and emphasised the need to maintain focus on the mixed type of warfare and to make irregular war a core competency.[48] The U.S. Army Special Operations Command White Paper on Counter Unconventional Warfare (CUW), while analysing capabilities of Russia, Iran and China, defines hybrid warfare as a combination of a conventional, irregular and asymmetric means to include the constant manipulation of political and ideological conflict.[49]

In January 2017, the International Security Advisory Board (ISAB), under the Department of State came out with a report on the 'Grey Zone Conflict'.[50] The report defined 'Grey Zone' as the use of techniques to achieve a nation's goals and frustrate rivals by employing instruments of power, often asymmetric and ambiguous in character, that are not direct use of acknowledged regular military forces. It characterised 'Grey Zone' as ambiguity about the nature of the conflict, opacity of the parties involved and in terms of violence fall between traditional war and peace. The Grey Zone techniques given in the document cover a large spectrum as given in Table 1.1: [51]

Table 1.1: Grey Zone Techniques as Defined in the ISAB Report 2017

- Cyber and information operations to undermine public/allied/local/regional resistance and information propaganda in support of other hybrid instruments
- Covert operations under state control, espionage, infiltration and subversion
- Special Operations Forces (SOF) and other state-controlled armed units and unacknowledged military personnel
- Support—logistical, political and financial—for insurgent and terrorist organisations
- Assistance to irregular or para military forces
- Economic pressure that go beyond the normal economic competition
- Manipulation and discrediting of the democratic institutions including electoral system and judiciary
- Enlistment of non-governmental actors, including organised criminal groups, terrorist, extremist political, religious, ethnic or sectarian organisations
- Calculated ambiguity use of/ covert/unacknowledged operations and deception and denial
- Explicit or implicit threat use, or threats of use of armed force, terrorism and abuse of civilian populations and of escalation

The increasing scope of definition of irregular or hybrid war from 2005 to 2017 in the U.S. military and State Department reflects the evolving character of hybrid warfare. While concluding, the ISAB report concurs that,[52] in future grey zone threats are likely to be more common and potentially more dangerous, including a nexus between criminal organisations and terrorist groups.

Perspective from China on Hybrid Warfare: The Three-Warfare Strategy. The first major shift in Chinese military strategy was in 1980 when, under the threat of invasion from the USSR, Deng Xiaoping adopted the strategy of 'Active Defence'. This was a major shift from the Mao strategy of fighting a protracted war deep inside Chinese territory.[53] Historically, ancient Chinese military literature from the time of Sun Tzu has laid more importance on hybrid warfare than conventional war, as has been discussed earlier in this chapter. The Communist Party of China (CPC) and the Central Military Commission (CMC) approved the concept of the 'Three Warfares' (San Zhongzhanfa) in 2003. This is a conceptual umbrella for three overlapping strategies of information warfare. They are:

- Coordinated use of strategic psychological operations.
- Overt and covert media manipulation.
- Legal warfare designed to manipulate strategies, defence policies, and perceptions of target audiences abroad.

These strategies empowered the People's Liberation Army (PLA) to conduct a function for 'political work operations' through 'public opinion', and 'psychological and legal warfare'. These three tools evolved to be the most lethal and effective tools of hybrid warfare. While Mao has been the doyen of '*Shi*' and 'guerrilla warfare', in modern China it was only in December 2003 that China as a policy endorsed applying public opinion, psychological and legal warfare as a strategy connected to hybrid warfare. The three-warfare strategy apparently appears to be non-kinetic and non-violent. However, during peace time, the three-warfare strategy will be the key to shape the battlefield (more about the application part has been covered in Chapter 4). A number of scholars connect PLA adopting the 'Three-Warfares' strategy with work of two PLA Colonels, i.e., '*Unrestricted Warfare*' published in 1999 In which Col. Qiao Liang and Wang Xiangsui studied the impact of globalisation and the resultant effect of merging of cultures, diplomacy, religion, politics and economics under the umbrella of technology. They concluded "*...in warfare and non-military warfare, which is primarily national and supra national, there is no territory which cannot be surpassed, there is no means* *that cannot be used, and there is no territory and method that cannot be used in combination.*" (sic).[54] The conceptual idea of 'Unrestricted Warfare', i.e., '*first rule of unrestricted warfare is that, there is no rule, nothing is forbidden*'[55] is what China has been pursuing in the South China Sea (the Chinese hybrid warfare in South China Sea has been covered in detail in Chapter 4).

> *In warfare and non-military warfare, which is primarily national and supra national, there is no territory which cannot be surpassed, there is no means that cannot be used, and there is no territory and method that cannot be used in combination.*
>
> **—Col Qiao Liang and Wang Xiangsui**

Chinese military strategy has evolved post-Gulf War 1991 which also kick-started the transformation of the PLA. The White Paper of 2015 with title of 'Military Strategy,' signalled a more coercive phase of China's rise.[56] The paper also covered the new security concept of 'holistic' security, which was introduced for the first time by President Xi Jinping at the Politburo study session in 2014. The new concept integrates both domestic and international security. The paper enlarges the security spectrum to include 11 fields—political, territorial, military, economic, cultural, social, science, and technological, information, ecological, financial and nuclear. An analysis of the recent Chinese reactions to various situations with the neighbours, at Doklam with India, at Senkaku Islands with Japan, Scarborough Reef with the Philippines and in the South China Sea that the 'holistic security' concept facilitates China to take advantage of disputes.[57] Military analysts in China believe that a military crisis handled properly provides a major opportunity to further national interest.[58] The creation of the Strategic Support Force (SSF) as a part of major reforms introduced in 2015 is an important milestone in the transformation of the PLA. These reforms are aligned to the 'Holistic Security Policy' as given out in the White Paper on Defence 2015, wherein the PLA seeks to expand its role from land-based territorial defence to power projection to protect Chinese interests in strategic frontiers of space, cyber space and the far seas.[59] The creation of the SSF centralizes the space, cyber, electronic and psychological warfare capabilities which were under various services and departments of the PLA under one HQ.[60] The Chinese white papers on defence have been analysed to study the evolution of Chinese military strategy and transformation of the PLA and its convergence with hybrid warfare in Chapter 4.

UK Prepares for Counter Hybrid Warfare. The Russian annexation of Abakhazia and South Ossetia in Georgia in 2008, Crimea in 2015 and the ongoing interference in Ukraine have caused concern in Europe. Europe and the UK which considered themselves relatively insulated from hybrid or sub-conventional warfare are now suddenly vulnerable. The Ministry of Defence of the UK initiated a project on 'Countering the Hybrid Threat' under the Multinational Capability Development Campaign (MCDC) in 2016-17. The MCDC has come up with two documents' the first is 'Understanding the Hybrid Warfare' and the second is on 'Countering the Hybrid Warfare'.[61] The project has worked out a baseline definition of hybrid warfare as '*the synchronized use of multiple instruments of power tailored to specific vulnerabilities across the full spectrum of societal functions to achieve synergistic effects*'.(sic)[62] The project report describes how an actor engaging in hybrid warfare may use wide-ranging instruments of power like military, political, economic, civil and information (MPECI) and aim to exploit the target the vulnerabilities like political, military, economic, social, infrastructure and information (PMESII).[63] The Figure given below explains the instruments of power and target vulnerabilities.

Figure 1.2: MPECI instruments of power and PMESII target vulnerabilities

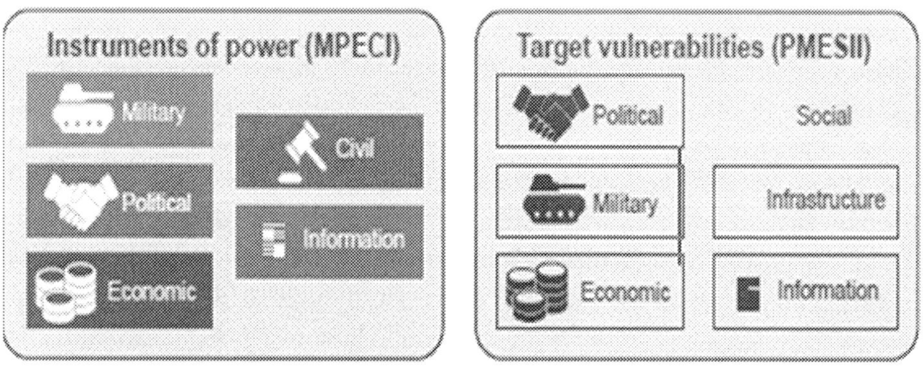

Source: MCDC Countering Hybrid Warfare Project Report, p. 13

Russian Perspective–Non Linear War. The Military Doctrine of the Russian Federation–2014, characterises the current military conflicts as integrated employment of military force and political, economic, informational or other non-military measures implemented with a wide use of the protest potential of the population and of special operations forces.[64] Under the heading of main military threats it says '*demonstration of military force in the course of exercises in the territories of states contiguous with the Russian Federation or its allies*'. Under the internal military risks the doctrine has identified the following threats:

- Activities aimed at changing by force the constitutional system of the Russian Federation; destabilizing domestic political and social situation in the country;
- Disrupting the functioning of state administration bodies, important state and military facilities, and the information infrastructure of the Russian Federation;
- Activities of terrorist organisations and individuals aimed at undermining the sovereignty and violating the unity and territorial integrity of the Russian Federation;
- Subversive information activities against the population, especially young citizens of the State, aimed at undermining historical, spiritual and patriotic traditions related to the defence of the Motherland;
- Provoking inter-ethnic and social tensions, extremism, stirring up ethnic and religious hatred or enmity.

It is quite interesting to note that the characterisation of the current military conflict and the internal risks identified lie in the domain of hybrid warfare. The mention of military exercises close or contiguous to Russian territory suggests the USA's interference in Ukraine or other countries along the border of the Russian Federation. Russia continues to perceive the USA as the primary threat.

The Russian 'Non Linear' concept of warfare is aligned to hybrid warfare. Gen Valery Grasimov, Chief of General Staff, published a paper in 2013 titled, '*The Value of Science is in the Foresight: New Challenges Demand Rethinking the Forms and Methods of Carrying out Combat Operations*'. The conceptual construct as given out by Gen. Grasimov describes nonlinear war as one which does not have distinct lines (forward/core/depth/rear) or friendly or enemy areas; it relies on the subversion and division of the enemy's social and political structure. It allows the aggressor to employ any means, not only brute force. The article by Grasimov illustrates employment of non-military to military measures as 4:1. The non-military means include social, diplomatic isolation, economic sanctions and religious disruptions. A graphical description of differences in the traditional employment of military forces and new form of war is given in tabular form in Figure 1.3.

The Russian actions in Georgia, Crimea, Estonia and Ukraine[65] in using propaganda, political and social subversions, protests and cyber-attacks show the implementation of non-linear/hybrid means. The above-mentioned operations have been the cause of military success in the last two decades for Russia and shows the current line of the Russian military.

Figure: 1.3: Graphic from Grasimov's article in *Voynno-Promyshlennyy Kurier*, 26 Feb 2013, translated by Charles Bartles[66]

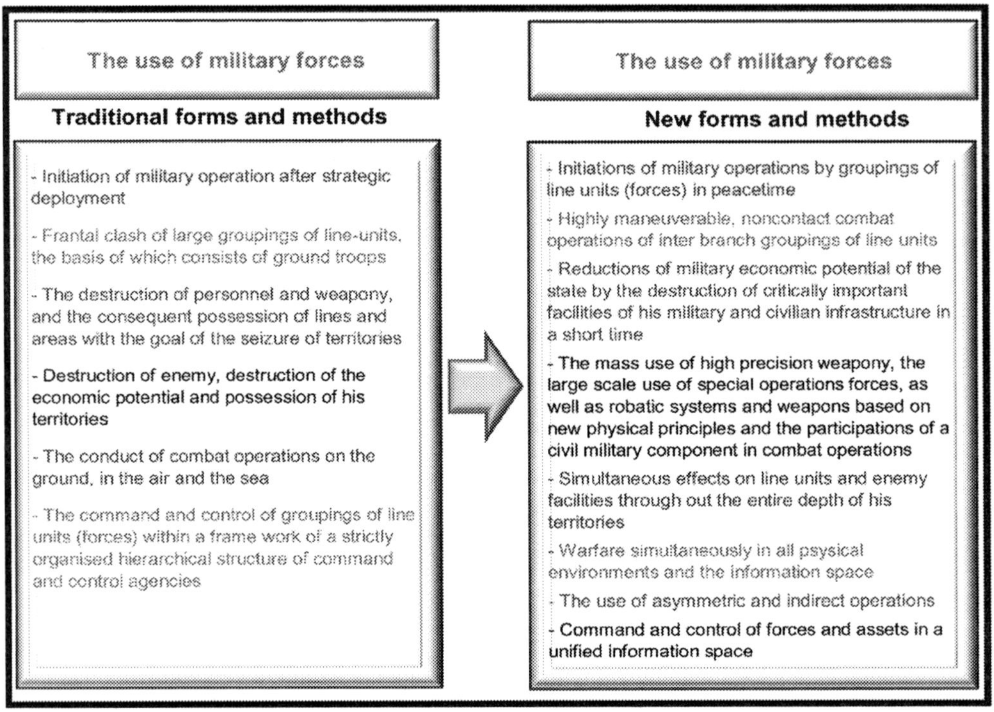

Indian Perspective. One of most comprehensive definitions of sub-conventional operations is given in the Sub-Conventional Doctrine of the Indian Army-2006. The sub-conventional operations arena is described in the doctrine as:

> '*Sub conventional operations arena comprises armed conflicts that are above the level of peaceful coexistence amongst states and below the threshold of war. These include militancy, insurgency, proxy war and terrorism either employed as part of an insurrectionist movement or independently. Border skirmishes also fall within this category.*'[67]
>
> **—Indian Army 'Sub Conventional Operations Doctrine-2006'**

The Indian Army Sub-Conventional Operations Doctrine-2006 was the outcome of more than 50 years of combating sub-conventional war. It aptly identifies the population as the Centre of Gravity (CoG), which changes the rules of engagement from conventional war, hence the training philosophy also undergoes a change. The doctrine also visualised that the Indian army is likely to remain engaged in sub-conventional operations for a long time to come. In the doctrine, the sub-conventional operations have been defined as:

"The sub-conventional operations are defined as asymmetric war which entails unconventional ways and means to target various elements of national power, to include command, control, communications, computers, intelligence, interoperability, surveillance and reconnaissance infrastructure, services-based industry, banking and financial services, transport and power sector. This type of warfare also includes cyber and information operations and border skirmishes."[69]

—Sub-Conventional Operations

Figure 1.4: Sub-Conventional Operations Area

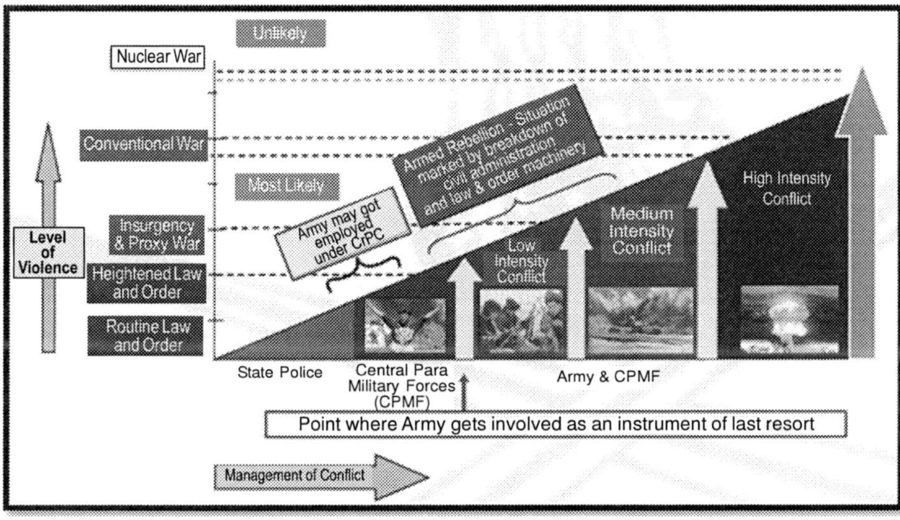

Source: Arena of Sub Conventional Operations as given in the Indian Army 'Sub Conventional Operations Doctrine-2006[68]

Figure 1.5: Sub-conventional Operations

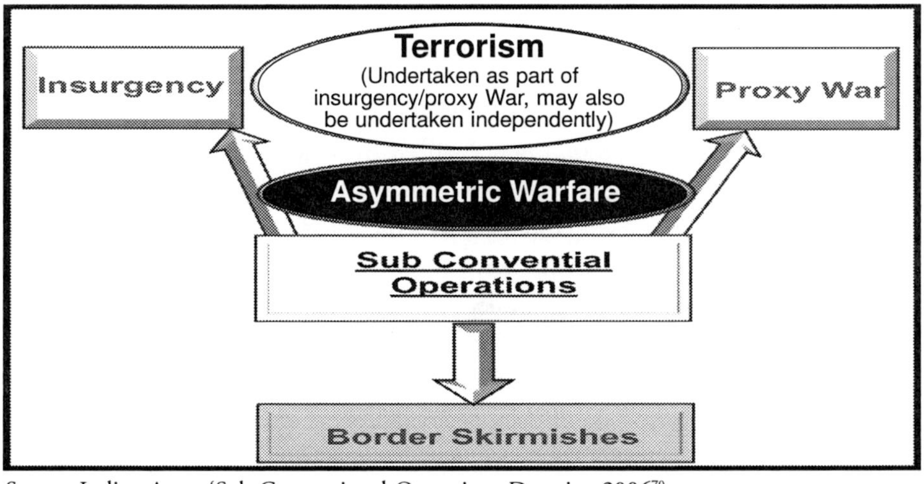

Source: Indian Army 'Sub Conventional Operations Doctrine-2006[70]

Indian Perspective on Hybrid Warfare. The Indian Army doctrine 2006 has given out the differences between conventional and sub-conventional war. It gives out the rule of engagements and highlights the importance of the challenges in the junior leadership, no violation of human rights, use of minimum force, perception management and interaction with the media. However, the aspects of cyber warfare and information warfare have not been covered. The threat of proliferation of agents of mass destruction, like improvised nuclear explosive device, radiological dispersal device (dirty bombs), chemical or biological agents to terrorists/non-state actors have been identified as future challenges. The conflicts post 2006 such as wars in Estonia, Georgia, Crimea, Lebanon, Yemen, Iraq and events in South and East China Seas have influenced the Indian perspective also. The Indian Army's Land Warfare Doctrine of 2018 under the future challenges identifies the battle space as a 'Grey Zone' and the character of warfare as 'Hybrid'. It states:

> *"Future conflicts will be characterised by operating in a zone of ambiguity where nations are neither at peace nor at war—a 'Grey Zone' which makes our task more complex. Wars will be hybrid in nature, a blend of conventional and unconventional, with the focus increasingly shifting to multi-domain warfare varying from non-contact to contact warfare."*
>
> **—Indian Army Land Warfare Doctrine-2018**

The Indian perspective has evolved over 70 years of combating regular/irregular/proxy war which is sponsored by Pakistan. The Chinese assertiveness on the Line of Actual Control (LAC), issues like Doklam and other standoffs on LAC, incidents of covert support to North East insurgent groups and the cyber warfare are also the factors which have influenced the Indian strategic thinking on Hybrid Warfare. The detailed analysis of Pakistan's and China's hybrid warfare has been discussed in chapter 3 & 4 respectively.

Hybrid Warfare: A Plural Perspective. The writings of Sun Tzu and Kautilya establish that the hybrid war conceptualized by scholars in the 21st century is similar to how it was in ancient times. Use of all available means of warfare in combination and with varying intensity to achieve the national aim was ingrained in the definition of war since ancient times. It was only during the Westphalia period or more so during the Napoleonic wars, which produced strategists like Clausewitz and Jomini, that the definition of war was compressed to the use of violence to achieve the national aim. Kautilya and Sun Tzu both lay higher emphasis on the use of intelligence, deception, decoys, subversion, economic gratification, making alliances and use of guerrilla techniques before the option of full-fledged war. The last 40 years or so have been one of the most violent periods of human history, where so many wars have been/are being fought, but not many have been direct wars between major/nuclear powers. Most of the wars fought have been indirect or wars of intervention or for regime change or against non-state actors. This violent period has led to revision/adaptation of the military doctrines of the major powers. The analysis of the doctrine or literature on perspective of hybrid warfare of the major powers bring out that all of them agree to the concept of employment of all means available to target the criticalities or vulnerabilities of target country. However, the nomenclature or classification may vary. The perception of each of the major powers has evolved because of environmental reality/challenges/threats specific to the country. The summary of the analysis of the doctrine/official literature on hybrid warfare of five major powers is given in Table 1.2.

Table 1.2: Perspectives of Major Military Powers of the World on Hybrid Warfare

Military Power	*Perspective on Hybrid Warfare*	*Influencing Factors*
USA	**Hybrid Warfare & Grey Zone Techniques.** As reflected in Quadrennial Defence Review 2006, National Defence Strategy 2008 and NSAB Report of 2017	• Experience in Iraq, Afghanistan, Syria • Impact of Russian operations in Georgia, Crimea, Ukraine • Chinese actions in the South China Sea
China	**Three-Warfare Strategy.** Three-warfare strategy in 2003 and creation of Strategic Support Force in transformation announced in 2015	• US-led operations in Iraq in 1992 & 2003 • Operations in Afghanistan • US initiatives in the Indo-Pacific • Transformation of technology
Russia	**Non-linear Warfare.** National Military Doctrine of 2014 and Non-linear war	• Transformation of technology • US interference in countries along the borders of the Russian Federation
UK & Europe	**Hybrid Warfare.** Ministry of Defence, MCDC report 2019, on countering hybrid war	• Impact of Russian actions in Georgia, Crimea, Ukraine and Estonia. • Increasing assertiveness of Russia
India	**Sub-conventional Warfare & Hybrid Warfare.** 2006 Sub-conventional doctrine and Land Warfare doctrine of 2108	• Pakistan-sponsored terrorism • Application of three-warfare strategy by China

Source: Compiled from inputs discussed in chapters 1, 3 and 4.

Framework for the Study. A framework was been prepared which identifies the criteria of classifying a conflict under hybrid warfare. The framework has been prepared based on the definitions given by scholars, experts and practitioners and the analysis of the doctrine and official publications of the major powers as discussed earlier in this chapter. In Chapter 2, the major conflicts of post-cold war will be analysed through this framework to establish the manifestation of hybrid warfare. Specific to India, the traditional adversaries, i.e., the actions of Pakistan and China will be analysed in detail in Chapters 3 and 4, respectively. The framework is explained as under and with a diagram in Figure 1.6:

Salient Characteristics of Hybrid Warfare. The concept given by Grasimov of three mutually overlapping principles has been taken as the base of the framework. These three aspects bring out the difference between conventional and hybrid war.

- **Permanency of Conflict.** The 'permanency of conflict' blurs boundaries between the states of war and peace, especially in terms of violence and means. The wars are traditionally declared and end with truce, surrender or signing of a treaty. However, in case of hybrid wars, the war need not be declared at all and it may continue after the violent phase is over. The presence of violence may or may not be there. This also reduces the impact of the 'time and space' factor in the hybrid warfare wherein a conflict may be over in a matter of a few days or may continue for decades. The conventional wars are traditionally fought amongst the nation states; however post-Cold War, the emergence of transnational organisations are launching wars against developed nations.
- **Multidimensionality.** The use of military or conventional means are no more the principal instruments of achieving the national aim. Rather it is the synergised application of traditional (kinetic) and non-traditional (non-kinetic) means to achieve the desired strategic result. By using multidimensional means apart from physical infrastructure, the cognitive domain of the adversary is also targeted. The various non-traditional and non-kinetic instruments are employed in hybrid warfare as given below:
  - Political Means
  - Cyber
  - Economic
  - Psychological
  - Informational
  - Use of irregulars or terrorists
  - Legal means.
- **Simultaneity.** In hybrid warfare, all the instruments available to the

nation are applied in a synchronised manner across the entire spectrum and depth of the physical domain of the adversary. The psychological and cognitive domains of the adversary along with the economic and social infrastructure are targeted by non-kinetic means while kinetic means would be used to target borders and vital areas in-depth simultaneously in phases.

Framework. Each hybrid operation will have its own signature and characteristic, which will be evident as case studies are analysed in the next three chapters. The Figure given below depicts a hypothetical hybrid operation by a transnational terrorist organisation against a state. The application of instruments of national power across various domains has been explained through this hypothetical example. In the left column, the three parallel coloured bars indicate the domains which are being targeted by the attacker. At the bottom of the model, the instruments of the national power are listed, which are also aligned to the categorisation of conflicts depending on the intensity and violence. In this particular case, the transnational terrorist organisation used propaganda, social and print media, and other instruments of psychology to target the population

Figure 1.6: Model depicting the employment of various instruments of a nation's power across various domains

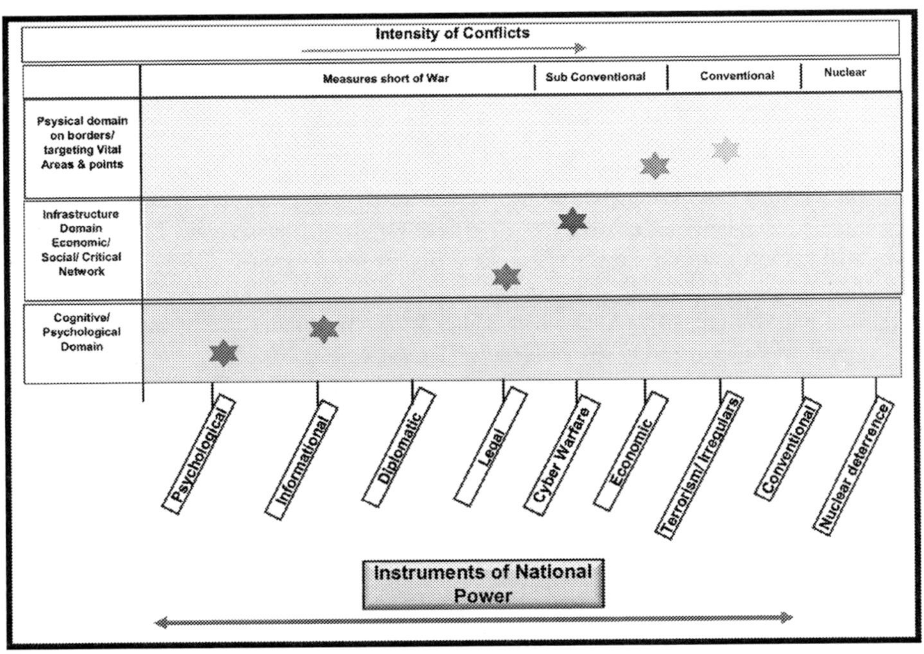

and the cognitive domain. It used cyber to attack the financial network. It captured the oil refinery and other profit-making organisations and started its own business to make the organisation self-sustaining in financial terms. Simultaneously it started orchestrating terrorist attacks in sensitive areas and points. All these activities can be marked on the model to create a big picture of the terror operations. The model can be further modified for the specific instruments being used.

Conclusion

The increasing globalisation, technological advancement in communication and modes of transport have transformed the emerging security challenges. Wars and conflicts are no longer restricted to conventional interstate rivalries based on territorial or ideological issues, but are driven by economic, demographic, societal and ethno-religious tensions that are transnational in nature.[71] As evident from Table 1.1, all major military powers recognise the need to address hybrid warfare as a separate issue in the science of warfare. The perspective of nations having evolved according to the threat they face led to the concept of plurality in its acceptance. Hybrid warfare not only needs specialised training, it also needs the organisation to be more flexible and responsive. Intangible aspects like intelligence, situational awareness, psychological preparedness and joint activities not only at the inter-service level but also at the national level assume importance. The national security strategy or military doctrines need to address the issue of hybrid warfare or whatever acceptable nomenclature is, so that due importance is given and nations are better prepared to face the threat of hybrid warfare.

NOTES

1. Indian Army Land Warfare Doctrine
2. Mary Kaldor, in review essay, *In Defence of New Wars*, published in *Stabilit* (2(1): 4, pp. 1-16, available at http://dx.doi.org/10.5334/sta.at, accessed on 04 January 2019
3. Ibid
4. Entienne G. Kurg; Linda L. Dhalberg; James A. Mercy; Anthony B. Zwi and Rafael Lozano (eds.), *World report on Violence and Health*, Geneva: WHO, 2002
5. Brig Gurmeet Kanwal, in *Hybrid Warfare The Changing Character of Conflict*, Pentagon Press, p. 6
6. Mary Kaldor, *In Defence of New Wars*, published in *Stabilit*, (2(1): 4, pp. 1-16 available at http://dx.doi.org/10.5334/sta.at accessed on 4 January 2019
7. Dr. Chris Tuck in *Hybrid War the Perfect Enemy* available at https://defenceindepth.co/2017/04/25/hybrid-war-the-perfect-enemy/ accessed on 4 Jan 2019
8. Ibid
9. Ibid

10. Carl Von Clausewitz '*On War*' translated by Michael Howard and Peter Paret, Oxford University Press, 2016, pp. XVI
11. Carl Von Clausewitz '*On War*' translated by Michael Howard and Peter Paret, Oxford University Press, 2016, pp. 12
12. Sun Tzu, *The Art of War*, translated by S. B. Griffith, Oxford: Clarendon Press, 1964
13. Kautilya *The Arthashastra* translated by L. N. Rangarajan, published by Penguin Classics, 1992, pp. 636
14. Sun Tzu, *Art of War*, translated by Lionel Guiles, Pax Librorum Publishing house, 2009, pp. 2
15. Ibid
16. Carl von Clausewitz "*On War*' translated by Michael Howard and Peter Paret, Oxford University Press, 2016, p. XXVI
17. Ibid
18. Ibid
19. Ibid
20. Napoleon Bonaparte's Peak of Military Success Ulm and Austerlitz at http://www.inquiriesjournal.com /articles/697/ accessed on 18 Dec 2018
21. Carl von Clausewitz "*On War*' translated by Michael Howard and Peter Paret, Oxford University Press, 2016, p. XXVI
22. Ibid
23. Carl von Clausewitz "*On War*' translated by Michael Howard and Peter Paret, Oxford University Press, 2016, p, XXVI
24. Mary Kaldor, in review essay, *In Defence of New Wars*, Published in *Stabilit* (2(1):4, pp, 1-16 available at http://dx.doi.org/10.5334/sta.at accessed on 4 January 2019
25. Ibid
26. Carl von Clausewitz "*On War*' translated by Michael Howard and Peter Paret, Oxford University Press, 2016, p. XXVIII
27. Kautilya '*The Arthashastra*' translated by L. N. Rangarajan, published by Penguin Classics, 1992 p. 636
28. Martin van Creveled in '*More on War*' Oxford University Press, 2017
29. Ibid
30. Sun Tsu, *Art of War*, translated by Lionel Guiles, Pax Librorum Publishing house, 2009, p. 7
31. Mary Kaldor, in review essay, *In Defence of New Wars*, Published in *Stabilit* (2(1):4, pp. 1-16, available at http://dx.doi.org/10.5334/sta.at accessed on 4 January 2019
32. Vikrant Deshpande and Shrabani Mehta, in *Contextualising Hybrid Warfare*, IDSA Publication, Pentagon Press, p. 33
33. Kautilya. *The Arthashastra* translated by L. N. Rangarajan, published by Penguin Classics, Fig. 1, p. XII
34. Mary Kaldor, in review essay, *In Defence of New Wars*, Published in *Stabilit*, (2(1):4, pp. 1-16 available at http://dx.doi.org/10.5334/sta.at, accessed on 4 January 2019
35. Martin van Creveld in *Transformation of War*, The Free Press, 1991, p. 8
36. Less Russia-China border skirmish of 1969; the incident cannot be categorised as a war as none of the parties declared it as such.(add more amplification in foot notes)
37. Ibid
38. Ibid
39. Brig Gurmeet Kanwal, in *Hybrid Warfare The Changing Character of Conflict*, Pentagon Press,

p. 4

40. Lt Gen (Retd.) A. K. Singh, in *Conventional Wars in Future War; Changing Nature of Conflict*, p. 60
41. Mary Kaldor, in review essay, in *Defence of New Wars*, Published in *Stabilit* (2(1),:4, pp.1-16, available at http://dx.doi.org/10.5334/sta.at accessed on 4 January 2019
42. Vikrant Deshpande in *Contextualising Hybrid Warfare, Hybrid Warfare: Changing Character of Conflict*, IDSA publication, Pentagon Press
43. Frank Hoffman in *Conflict in 21st Century: Rise of Hybrid Warfare*, Potomac Institute of Policy Studies, Arlington, Virginia, 2007
44. The case studies of Georgia and Crimea have been covered in chapter 2.
45. *Quadrennial Defence Review* (QDR)-2006, U.S. Department of Defence available at https://archive.defense.gov/home/features/2009/0509_qdr/index.html accessed on 30 May 2019
46. Ibid
47. *National Security Strategy 2008*, U.S. Department of Defence, available at https://archive.defense.gov/home/features/2009/0509_qdr/index.html accessed on 30 May 2109
48. Gen Mattis, in *US Must Prepare for Hybrid Warfare, Small Wall Journal*, 2009, available at https://smallwarsjournal.com/blog/general-mattis-us-must-prepare-for-hybrid-warfare accessed on 30 May 2019
49. Special Operations Command White Paper on Counter Unconventional warfare-2014 available at https://info.publicintelligence.net/USASOC-CounterUnconventionalWarfare.pdf accessed on 1 Jun 2019
50. DoS, IASB report on *Grey Zone Warfare*, 2017, available at https://findit.state.gov/search?query=grey+zone&affiliate=dos_2009_2017_stategov&search= accessed on 30 May 2019
51. Ibid
52. Ibid
53. M Taylor Fravel in Active Defence- China's Military strategy since 1949, Princeton University Press, Princeton and Oxford, 2019
54. Col Qiao Liang and Wang Xiangsui in Unrestricted Warfare, Beijing, : PLA Arts and Literature Publishing House, 1999, p 162
55. Ibid
56. The Chinese white papers on Defence have been analysed in Chapter 4.
57. These incidents have been analysed as case studies in chapter 4.
58. Zhao Zijin and Zhao Jinfang, 'on the Control and Management of Military Crisis" China Military Science, Jul 2013 pp 62-71 quoted by Timothy R. Heath, in The 'Holistic Security Concept': The Securitization of Policy and Increasing Risk of Militarized Crisis, in China Brief, in James Town China Brief volume 13 issue 9, available at https://jamestown.org/program/the-2013-defense-white-paper-in-perspective/ accessed on 02 Feb 19
59. David Blasko in PLA as Organisation V 2.0, in PLA Ground Forces, p. 262.
60. Jhon Costello and Joe McReynolds in China's Strategic Support Force, in China Strategic Perspectives, No 13 published by Centre for the Study of Chinese Military Affairs Institute for National Strategic Studies, National Defence University Press, Washington
61. MCDC Countering Hybrid Warfare Project Reports available at https://www.gov.uk/government/publications/countering-hybrid-warfare-project-understanding-hybrid-warfare accessed between 5 Mar to 30 May 2109

62. MCDC Countering Hybrid Warfare Project Report on 'Understanding Hybrid Warfare' available at https://www.gov.uk/government/publications/countering-hybrid-warfare-project-understanding-hybrid-warfare accessed between 5 Mar and 30 May 2109
63. Ibid
64. Military Doctrine of Russian Federation-2014, available at https://rusemb.org.uk/press/2029, accessed on 30 May 2019
65. The case studies on Georgia, Crimea, Estonia and Ukraine will be discussed in Chapter 2
66. Charles K. Bartels in *Getting Grasimov Right*, available at https://www.researchgate.net/publication/329933852_Getting_Gerasimov_Right, accessed on 3 Jun 2019
67. Indian Army Sub-conventional Doctrine 2006, available at http://indianstrategicknowledge online.com /web/ doctrine%20sub%20conv%20w.pdf accessed on 28 May 2019
68. Ibid
69. Indian Army Sub-conventional Doctrine 2006, available at http://indianstrategic knowledgeonline.com /web/ doctrine%20sub%20conv%20w.pdf, accessed on 28 May 2019
70. Ibid
71. Gurmeet Kanwal and Samoset Ghost in Future Wars Changing Nature of Conflict, Matrix Publications, p. xi

2

Transformation of Hybrid Warfare and Emerging Trends

"Militaries are today faced with a choice: they can stick to what they know and do best, at the risk of reduced relevance. Or else they can re-order themselves to deal with the new challenges that face us, rethinking doctrines and practices from the tactical to the operational to the strategic and even the grand strategic level."

—Shiv Shankar Menon, Former NSA,
(Address, FM Cariappa Memorial Lecture, 5 Oct 2011)

ABSTRACT

This chapter brings out the transformation of hybrid warfare through case studies of conflicts since the post-cold war. The emerging character and future trends in hybrid war including the impact of disruptive technologies have been covered. A retrospective analysis of operations in the Eastern Theatre in 1971 through the prism of hybrid warfare has been carried out.

The Prussian military philosopher, Carl Von Clausewitz, in his seminal work, 'On War' says, "War is more than a true chameleon that slightly adapts its characteristics to the given case"[1] He further characterises war as a "remarkable trinity" involving or comprising violence or hatred, chance or probability and political consideration—that play out through the interaction and interplay of people, military forces and government. These factors have been integral to the character of war since the historic times of Sun Tzu and Kautilya as discussed in Chapter 1. Similarly, Clausewitz stated nearly two centuries ago that, '*although war changes its characteristics in various circumstances, in whatever way it manifests itself, war is still war.*' War since the twentieth century has evolved at a very rapid pace, making it difficult for philosophers or scholars to analyse and predict the future course. The traditional concepts of war came into prominence post 1648, i.e., the Westphalian concept, wherein States go to war to preserve their national interest.[2] The Napoleonic wars and writings of two great philosophers, Carl von Clausewitz and Jomini, who fought as part of Napoleon's Army at some point of time, further reinforced the point that war is fought and restrained for reasons that are rational and political in nature.[3]

The advent of nuclear weapons and the possible consequences of a major war between the nuclear powers brought out the importance of deterrence. However, this was the period when concepts like measures short of war or indirect war between the nuclear powers evolved. A number of little wars were fought between smaller nations under the respective umbrellas of NATO and Warsaw Pact. Martin Van Creveld in his book '*The Transformation of War*' argues that nuclear weapons were built to give the military and the nation states unprecedented powerful tools for making and winning the war, however, the nuclear weapons have actually put an end to the possibility of war between two nuclear armed nations.[4] He argued that not only the nuclear armed states have not gone for a war since 1945, even the allies who have 'protection' have been immune to conventional war.[5] The most important wars since 1945 have been between non-nuclear powers, the Middle East (1948-1949, 1952, 1967, 1973, 1982 and 1980-1988), between China and Taiwan (1954, 1958), the USA and Vietnam (1964), India and China (1962) and India and Pakistan (1947, 1965 and 1971).[6] Martin von Creveld has also argued that future wars will not be fought between the states in Clausewitzian style but would be fought largely by non-state actors, guerrillas, proxies and terrorists.[7]

> *In the increasingly globalised world, the emerging security challenges are no longer defined or restricted to conventional inter-state rivalries, but are cumulative outcomes of economic, demographic and social tensions.*
>
> —**Brig Gurmeet Kanwal**

Brig. Gurmeet Kanwal, writing on the changing character and taxonomy of conflict, brings out that in the increasingly globalised world, the emerging security challenges are no longer defined or restricted to conventional inter-state rivalries, but are cumulative outcomes of economic, demographic and social tensions.[8] In the globalised world and with modern means of travel and communications, these challenges more often evolved into transnational threat. Mary Caldor and Brig. Gurmeet, in their respective essays on the subject, have brought out that there has been a marked decrease in inter-state wars, while the number of civil wars have been increasing since WWII.[9, 10] The binary concept of war and peace started losing relevance the post-cold war era. War in the 21st century traversed into unfamiliar forms that employ both conventional and irregular forces on the same battlefield.[11] The nation-states are exercising more and more non kinetic and unconventional strategies which are much beyond the traditional military means and employ the complete range of instruments of national power like political and economic coercion, targeted assassination, sponsoring of irregulars and transnational terrorist groups, use of traditional and social media to carry out propaganda and offensive cyber operations. This emerging change in the character of war from being rational to irrational and from use of conventional means to simultaneous employment of a combination of all means of national power has led to the manifestation of hybrid warfare. This chapter will analyse the transformation in the character of war post the end of the cold war period till date to highlight the manifestation of hybrid warfare. The chapter will also focus on the likely futuristic trends in the hybrid warfare. The case studies of Pakistan–India, Pakistan-Afghanistan and China's hybrid warfare have been covered separately in Chapters 4 and 5. However, the pattern and characteristics of those case studies have been used at the end of this chapter for the analysis part.

Israel-Hezbollah War-2006

The Israel-Hezbollah War of 2006 is a turning point in the history of warfare. Before this war, Israel had won almost all the wars fought against Arab nationalism, at times while fighting simultaneously with more than one Arab state. But in 2006, Israel was not fighting an Arab state, it was fighting a 'state within a state', or 'a tribal faction' or a 'sect of religion' or a 'transnational terrorist organisation'. Technically Israel did not lose the war against the Hezbollah in pure mathematical terms, but it did not win it either.[12] However, if weighed against the parameter of asymmetry that Israel enjoyed in terms of conventional forces against the Hezbollah, then possibly Israel lost, it lost to a terrorist organisation fighting a hybrid war. The trigger for the war was the killing of three and kidnapping of two Israeli soldiers by the Hezbollah on July 12, 2006.[13] The terrorist organisation never expected the way Israel retaliated. The Israel air force launched a merciless massive air campaign in South Lebanon. In the initial phase of the war, the Israeli air force carried out massive bombing operations against the Hezbollah's stockpile of Katyusha third-generation rockets and also claimed to have destroyed the HQs of the Hezbollah near Beirut.[14] The Israel ground forces supported with a complete spectrum of conventional means entered South Lebanon on 17 July. It was in the fight for villages in South Lebanon where Israeli mechanised forces were not only surprised but shocked. The Hezbollah was able to integrate the local villagers into their battle plan, the fighting positions were well-sited, defended, and stocked with ammunition as well as food supply.[15] Israeli intelligence had failed in their estimation of Hezbollah's preparation for war. Hezbollah fighters operated in small teams, providing decentralisation flexibility and initiative.[16] They tactfully used the houses as anti-tank weapon positions and effectively used a wide range of anti-tank launchers and missiles. In one particular tactical encounter, one Israeli tank column was halted for two days and more than 11 tanks destroyed.[17] In spite of the Israeli air force claiming to have destroyed the stockpile of rockets, the Hezbollah succeed in preserving a good stock in the villages and civil area. These Katyusha third-generation rockets created havoc in the daily life of the normal Israeli people and severely affected the economic activities in Northern Israel.[18] The Hezbollah commander, Nasrallah, had adapted guerrilla warfare in the true sense. Hezbollah fighters were hiding or had merged with the local population hiding in homes, schools, farms, cattle sheds, churches and mosques. He was successful in converting the battle into a war between Israel and pan Arab nationalism, the invincibility of the Israeli army was put under a question mark.[19]

Another area where the Hezbollah outpaced the Israeli defence forces was the information war. The Hezbollah commanders understood the nuances of the revolution that was taking place in the domain of information warfare.[20] The transformation of a journalist from mere objective reporting of happenings on the battle field to an opinion maker or an activist because of 24/7 TV news channels was exploited by the Hezbollah.[21] The journalists were taken to the battle zone, they were shown where and how the Israeli air force and artillery had bombed the civil areas. The damage, suffering and loss of civil life especially to women and children due to Israeli action, was the focus of the media campaign.[22] The Hezbollah strategy of handling the media paid off. Soon the narrative in the media, including the Western media, was the 'disproportionality' and 'irrationality' of Israeli military action. The media was toeing the line that while the Hezbollah may have triggered the crisis, the air raids and artillery firing destroying the infrastructure, targeting of hospitals and schools, killing of women and children of Lebanon as tit-for-tat action for the abduction of just three Israeli soldiers and was not justified by any parameter.[23] The Hezbollah armed and supported by pro-Arab channels like Al Jazeera and Al-Arabiya took the destruction caused by the Israeli air campaign into every sitting room and bedroom not only in the Arab region but also to other continents.[24] The effectiveness of the Hezbollah's information war was also reflected in the reporting pattern of Western media giants like the *BBC*, *Washington Post* and *New York Times*. The *BBC* ran 117 stories on the war, of which 30 percent showed Israel as the aggressor and only 3 per cent showed the Hezbollah as the aggressor.[25] The front page headlines in the *New York Times* and *Washington Post* reported Israel as the aggressor as much as two times more than they did for the Hezbollah.[26] A normal viewer was made to believe that Israel was the aggressor and that Lebanon was the victim.

Marvin Kalb and Carol Saivetz, in their paper '*The Israeli–Hezbollah War of 2006: The Media as a Weapon in Asymmetrical Conflict*' have quoted the famous *BBC* anchor Nik Gowing who, while at a panel discussion at Harvard University, said, "*new asymmetric information—the new level of accountability and public perceptions in a time of crisis exposed the vulnerability of traditional institutions of power and influence.*"[27] The traditional institutions like the Israeli defence forces are more and more vulnerable to asymmetric information which can have a detrimental effect on the public perceptions like a short visual footage of a Hezbollah fighter hitting an Israeli Merkawa tank with a hand-held rocket launcher had boosted the morale of every villager in North Lebanon who was supporting the Hezbollah. A similar incident in Jammu and Kashmir, when a young officer of the Indian Army preferred a nonviolent action to save the members of the Election Commission of India who were on duty to carry out work related to the elections were surrounded by a mob and an onslaught of stones. The officer could have resorted to firing for self-defence and saving the lives of the government officials. However, he decided to pick one of the stone pelters and used him as a shield to move the party of government officials safely out.[28] However, the vernacular media armed with the visual footage generated the narrative against that young Indian Army officer. The public was made to forget human rights were equally applicable to those Election Commission officials who were performing their duty. While maximum time the impact of such information bytes may be for a limited period, however, for that limited period the impact that such bytes may generate can have an asymmetric impact on public perception. The Hezbollah leadership understood this nuance well whereas a traditional institution like Israeli Defence Forces (IDF) preferred to continue with the traditional means of forcing censorship during times of conflict.[29]

In 2006, while Israel may have claimed to have destroyed the fighting capability of the Hezbollah significantly, however when compared with the performance of armies of other Arab countries in the previous Arab-Israel wars, the Hezbollah was not only able to stall the progress of the IDF but also caused damage.[30] The success of the Hezbollah can be attributed primarily to two factors; the first being the change in tactics—the guerrilla warfare and second one being better utilisation of information warfare means.

Cyber Attack on Estonia – 2007

The cyber-attack by Russia on Estonia in April 2007 lasted for approximately 22 days. In the conflict there was no violence or kinetic action between the two nations. The conflict was part of the larger political struggle wherein Estonia saw Russia as an oppressor.[31] Russia targeted Estonia only by cyber warfare and crippled the nation. The study of this conflict facilitates understanding of the nonlinear warfare being implemented by Russia. Estonia was part of the USSR since 1940. The ethnic Russians were in a minority in Estonia and were about 26 percent of the total population. Post the disintegration of the USSR, Estonia became an independent country. The relations with Russia started deteriorating when Estonia went on to join the EU and NATO.[32] The trigger of the conflict was connected to the decision of the Estonian government to relocate statues of the Russian Red Army soldiers who had died during WW II while defending Tallinn (current capital of Estonia). While the Russians connected with the two statues as liberators, the Estonians looked at them as oppressors. The work of relocating the statues began on April 26, 2007 when a peaceful gathering of pro-Russian protestors started growing and by evening the protests had turned violent. Cyber-attacks started by midnight of April 26-27, 2007.[33] Estonia is a highly digitised and networked country. Even during 2007 it was highly dependent on computer networks. All government offices were connected through networks and services were online including banking. Thus, complete governance and daily life was vulnerable to cyber-attacks.[34] The cyber-attacks continued till 18 May and in between peaked up on May 9, 2007. A larger composition of the attack can be described as Denial of Service (DoS) or Distributed Denial of Service (DDoS) attacks. Complex attacks were also recorded, which hacked into systems, for example using Structured Query Language (SQL) injection. The primary targeted entities included the government, the president, the parliament, police, banks, Internet service providers (ISPs), online media, as well as many small businesses and local government sites. The complete banking system was hacked, no ATM or bank could function. Web servers, e-mail servers, Domain Name System (DNS) servers and routers were the common targets. The highest number of attacks on May 9 may be linked to the historical and emotional importance of the day which the Russians celebrate as the victory day against the Nazis. The nation was brought down to its knees and the government was rendered helpless in just about 48 hours.

The study of cyber conflict between Russia and Estonia establishes the emerging vulnerabilities of a networked society. Though this was a relatively minor incident, it brings out the importance of Russia's nonlinear way of fighting.

Georgia – 2008

The 2008 conflict between Georgia and Russia involves the two Russian majority areas of South Ossetia and Abkhazia. These two enclaves are geographically part of Georgia but have been claiming independence from it. South Ossetia is approximately 1,500 square miles, with a population of only 55,000, most of whom live in Tskhinvali, the capital city.[35] Since 1921, South Ossetia has been an autonomous oblast of the Georgian Soviet Socialist Republic. After the collapse of the USSR, South Ossetia demanded the status of an independent republic, which was firmly denied by Georgia. This was followed by a war between Georgia in South Ossetia which was terminated in a ceasefire with the intervention of Russia. While, South Ossetia remained autonomous, a large part was under the forceful occupation of Georgia.[36] South Ossetia is sandwiched between Russia and Georgia. (refer Map). Abkhazia covers a territory of 8,660 square kilometres and has a population of 52,500 approximately. Abkhazia is surrounded by Russia on the north and west, on to east it shares a border with Georgia and on the south there is the Black Sea. (refer Figure 2.1). Abkhazia had been demanding the status of an independent republic which was denied by Georgia. In 1990-1991, there was a war between Georgia and Abkhazia and peace was brokered by Russia in 1992. South Ossetia and Abkhazia both had a presence of Russian peacekeepers till 2008 when attempts by Georgia to forcefully integrate the provinces prompted a strong Russian reaction and led to the Georgia-Russia war of 2008.

The Russia-Georgia War – 2008, is an important milestone in the history of warfare. It was for the first time that cyber and a conventional shooting war manifested simultaneously. Georgia had started gravitating towards joining the European Union (EU) and becoming part of NATO, which raised Russian concerns. The relations had been deteriorating after the 2003 'Rose Revolution'

Figure 2.1: Map Showing Abkhazia and South Ossetia

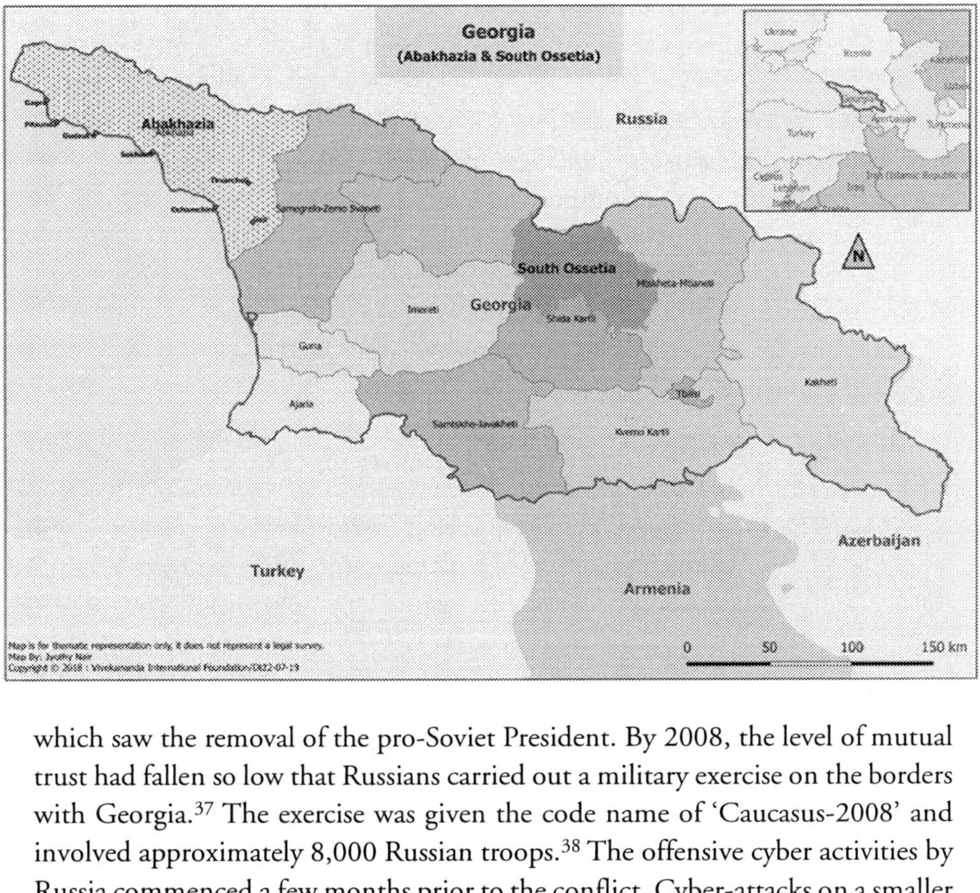

which saw the removal of the pro-Soviet President. By 2008, the level of mutual trust had fallen so low that Russians carried out a military exercise on the borders with Georgia.[37] The exercise was given the code name of 'Caucasus-2008' and involved approximately 8,000 Russian troops.[38] The offensive cyber activities by Russia commenced a few months prior to the conflict. Cyber-attacks on a smaller scale were being mounted on government sites and networks by Russian hackers since July 2008.[39] The cyber-attack carried out by Russia can be divided into two phases. In phase one, which started a month earlier, government sites and networks were targeted by hackers. The experience gained from Estonia was also used by Russian hackers, though the damage in Georgia was lesser as compared to Estonia because of limited penetration of the Internet and computer networks.[40] The second phase started on August 7-8, 2008 when Georgian ground forces moved into South Ossetia and Abkhazia. In this phase, critical and important networks like financial institutions, banking (ATMs), communications, business and transport were targeted. Till August 20 most of the major websites of the Georgian government were down, hampering the administration adversely. Georgia was cut off from the world. The Russian proxies in these two states supported the separatists to resist and revolt against Georgia, the tempo of protest and violence went up in July 2008, which forced the Georgian government to send forces into South Ossetia and Abkhazia on August 7, 2008. In response to Georgia's ground offensive, Russian ground forces entered South Ossetia and Abkhazia supported by fighters and artillery. The Russian air force achieved air superiority and the navy established a blockade in the Black Sea. By 10 August, Georgia withdrew its troops from South Ossetia and Abkhazia and the ceasefire was declared. On 25 August Russia recognised South Ossetia and Abkhazia as independent countries. Nicaragua too extended diplomatic recognition to these two countries. However, Russian efforts to get recognition to these two states in the Shanghai Cooperation Organisation (SCO) and Collective Security Treaty Organisation (CSTO) did not yield immediate results.[41]

Analysis of Georgia-Russia 2008 War. The 2008 war was another good example of how the Russians achieved their strategic aim with minimum effort and resources. The Russians needed to check Georgia's increasing engagement with NATO and the EU and simultaneously retain control over the states of South Ossetia and Abkhazia. They achieved this by first exploiting the vulnerabilities of Georgia in these two states through their proxies. The protest were so calibrated that the Georgian government was forced to use its forces to control the situation. This provided an opportunity to intervene legally and provide an apparent legitimacy to the Russian move. By using cyber capabilities, Russia ensured that Georgia was cut off from the world for the crucial period. Through attacks on critical infrastructure like finance, transport and communication networks, Russia conveyed the vulnerabilities to Georgia. By the time the Russian military moved in, Georgia had no option but to sign the ceasefire agreement on terms dictated by the Russians. Russian forces physically moved in but no major battle was fought between them and the Georgian army. Thereby, Russia was able to keep the situation well under the threshold level for other major/regional powers to intervene. The war in Georgia again reflects the Russian doctrine to use the entire spectrum of national instruments in a hybrid mode and in a nonlinear fashion to achieve a national aim.

Crimean War – 2014

Crimea is a strategically important peninsula on the north coast of the Black Sea in Eastern Europe. According to the 2001 census, Russians comprised 58.32 percent of the population, Ukrainians 24.32 percent, Crimean Tartars 12.1 percent,

Byelorussians 1.4 percent, Tartars 0.54 percent, Armenians 0.43 percent, Jews 0.22 percent and Greeks 0.15 percent.[42] Crimea became part of the USSR after the Soviet Revolution in 1917. In 1954, Crimea became part of Ukraine. In 1991, when the USSR collapsed, Ukraine became an independent country and the Crimean peninsula became an autonomous republic within Ukraine. The ethnic Russians of Crimea could never become part of Ukraine and suffered from a feeling of alienation.[43]. The Black Sea Naval Fleet of Russia is based at Crimea (Sevastopol). It provides strategic capability of force projection into the Black and Mediterranean seas.[44] Mr. Yushchenko, the country's former 'pro-Western' president of Ukraine, repeatedly threatened to expel Russia's Black Sea Fleet from Sevastopol. During his time, the EU and US interference in the functioning of Ukraine had increased. Certain other decisions like denial of autonomous status to Crimea by Kiev which was against the 1996 constitution, added to Russia's fears of increasing the US and NATO influence on Ukraine.[45] In 2014, the Ukraine parliament deposed President Yanukovych. The EU, NATO and the USA immediately approved the coup and recognised the new government in Kiev, whereas Russia did not issue any comment on the subject. The actions by the newly elected government in Kiev further alienated the Russians, especially in Crimea, like the Russian language being spoken by a majority was downgraded of its status as a regional language in 2014 by the new government.[46] The February 2014 revolution created a political crisis in Crimea.[47] Demonstrations broke out in Crimea against the new government in Ukraine; there were clashes between the ethnic Russians and supporters of the new government in Kiev. In Sevastopol, a parallel government was formed. Meanwhile, the Russian government deployed its military along the Ukrainian border in the garb of a military exercise.

The Russian government exploited the situation to achieve its long pending strategic aim, i.e., to get control over Crimea to have a secure base for its Black Sea fleet. The Night Wolves Motorcycle club,[48] which is based in Russia and is significantly funded by Russian government agencies, assisted the local Crimean protesters to organise self-defence units. Observers have claimed that Russian soldiers moved into Crimea in the garb of motorcycle riders. The group also assisted the protesters in setting up of check points, creating road blocks, spreading propaganda material and organising protests. The club played a leading role in creating a favourable situation for the Russian forces to move in by collecting intelligence and building public opinion. On February 25, 2014, thousands of pro-Russian Crimean protestors barricaded the Crimean Parliament and while waving the Russian flags called out for a referendum. The Russian army set up a check point between Sevastopol and Simferopol on the main highway connecting Crimea to Ukraine. (Figure 2.2). On February 27, men in green uniform without any insignias seized important buildings like the Crimean Parliament, radio station and council of ministers. Soon, the Russian flag was hoisted on these buildings. These green-uniformed men were later reported to be from the Russian Special Forces. The hacking of the internet and telecommunication networks ensured that Crimea was totally cut off from the rest of the world. The telecommunication networks of Ukraine were continuously attacked by Russians using the Crimean networks.[49] Russian naval vessels stationed at Sevastopol were involved in jamming Ukraine radio networks.[50] An emergency session of parliament was convened and Sergey Aksyonov, pro-Russian leader (Russian Unity Party) was elected as prime minister and a decision was taken to conduct the referendum on autonomy.[51] On March 1, 2014, the new prime minister formally asked President Putin for assistance to maintain law and order which made the legal way for the Russian military to move in. The date of referendum was advanced and on March 16, 2014, approximately 95 percent of the Crimean population voted to join Russia. So, Russia achieved its aim of securing control over the strategic Crimean peninsula in less than a month without using any kinetic means.

Analysis of Russian Strategy in Crimea. The Russian strategy can be broadly classified under the following heads. The means given below were used in a near-simultaneous timeframe, to achieve the necessary impact. The Ukrainian government was rendered helpless by Russia:

- **Exploitation of the Existing Fault Line.** Russians exploited the discord and disharmony between the ethnic Russians and Far Right Ukrainians in Crimean society.
- **Use of Proxies.** Russian agencies used the groups like the Night Wolves Motorcycle Club, which is known to be a staunchly nationalist group as its proxy. The group was brilliantly used as a force multiplier to carry out covert activities from intelligence gathering and carrying out propaganda to overt activities such as leading protests and organising mass demonstrations.
- **Employment of Non-uniformed Soldiers.** The 'little green men', were Russian soldiers without insignias. They were employed to take control of the power centres like parliament, radio stations and important communication nodes.
- **Information Operations.** Russian propaganda lasted for many years to create public opinion. At the crucial stage, the Night Wolves Motorcycle

Club was also used. Naval vessels of the Black Sea Fleet were used to jam the radio networks of Ukraine.

- **Cyber Warfare.** Russian agencies used the networks of Crimea to penetrate into the network of Ukraine and take control of the entire network. This completed the isolation of Crimea from Ukraine.
- **Use of Conventional Forces.** Russian military was deployed along the Ukraine border in the garb of carrying out an exercise. This was used to create psychological pressure on the Ukrainian government.

It is evident that the use of conventional forces was just symbolic and meant to convey a message; the major impact was achieved by non-kinetic means. That Russia achieved its aim of securing control of Crimea and the strategically important Sevastopol port for its Black Sea Fleet in just about a month of conflict without any violence is a classic example of hybrid warfare.

Figure 2.2: Map of Crimea

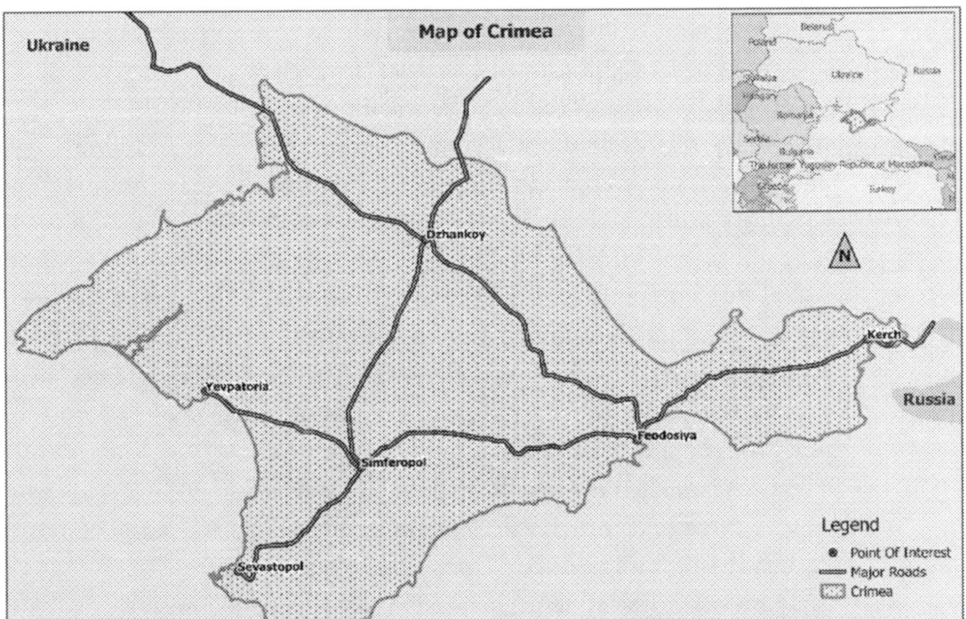

Emergence of ISIS: War in Syria and Iraq in 2014

The emergence of a transnational terrorist group like the Islamic State of Syria and Iraq (ISIS) has been one of the most defining events of the 21st century for scholars and academicians of war studies. This was the first time when a transnational terrorist group started a campaign to capture nation states to establish a new country based on its own ideology. Within a year, (July 2013 to July 2014) the ISIS controlled more than 34,000 square miles in Syria and Iraq and expanded it from the Mediterranean coast in the west to the south of Baghdad in the east. It had established its capital at Raqqa in Syria, was generating its own revenues and running the administration. The occupied territory was declared as a "Caliphate" by the group's leader, Abu Bakr al-Baghdadi, on December 14.[52] The meteoric rise was followed with an equally fast fall—by March 2019, the ISIS had lost all its strongholds. The case study of the ISIS brings out various facets of hybrid warfare. For a detailed analysis, the time line of important events is given below.

- In October 2004, Abu Musab al-Zarqawi forms Al-Qaeda in Iraq (AQI). While, AQI was also one of the sectarian groups fighting the coalition forces in Iraq at that time, its main agenda was to trigger a full civil war between the Sunni and Shia populations of Iraq. It carried out targeted attacks on population centres and the holy sites of Shias.[53]
- On 15 October 2005, the Al-Qaeda in Iraq (AQI) merged with other Sunni Jihadist factions to form the Islamic State in Iraq (ISI).
- In April 2010, Abu Bakr-al-Baghdadi became the leader of the ISI. He moved from Iraq to Syria in April 2011. He merged the Islamic State in Iraq (ISI) with the Nusra Front in Syria to create "The Islamic State in Iraq and Syria (ISIS)."
- In July 2013, the ISIS launched its campaign in Iraq and Syria to capture territory and by December 13, it took control of Fallujah and parts of Ramadi in Iraq. By January 2014, the ISIS captured Raqqa and declared it as its capital. By July 2104, it had control of approximately 3,400 sq miles in Iraq and Syria.
- The USA launched the first air strikes in Iraq against the ISIS advance in August 2014 and in Syria in September 2014. The US-led operations against the ISIS was named "Operation Inherent Resolve."
- 2015 saw the ISIS influence increasing globally; its followers carried out terror attacks in Paris, Egypt, Libya, Yemen, Tunisia, Saudi Arabia, Damascus, Beirut, Dhaka Bangladesh and Turkey and even in California. Affected countries like Egypt and France also joined the campaign with the USA against the ISIS.
- In 2016, the ground offensive by Iraqi and Syrian defence forces supported by the USA and the air forces of allied countries forced the ISIS to start

retreating. Between 2017 and 2018, the ISIS suffered a series of major defeats; it lost its major strongholds like Mosul and Raffa. On March 23, 2019, the ISIS lost its last stronghold in Syria and Syrian defence forces declared Syria free of the 'caliphate.'

The ISIS emerged from a small Sunni sectarian group called Al Qaeda in Iraq, which was primarily involved in carrying out targeted attacks on the Shiite community. Its transformation into a transnational terror group in a short period and the spectacular successes in the battlefields in Iraq and Syria took the entire world by surprise and shock. The strategy and tactics used by the ISIS can best be summarised as hybrid warfare. The blitzkrieg of the ISIS is a contemporary and relevant case study to understand hybrid warfare. The analysis of the hybrid means used by the ISIS is summarised as under:

- **Ideological Warfare.** The ISIS emerged and grew by exploiting the sentiments of the minority Sunni group against the majority group of Shiites.[54] It incited the Sunni population against the Shiite groups by targeting the holy places and community centres. It utilised the network of sympathisers to form the nucleus and built the organisation.[55]
- **Use of Conventional Means along with Guerrilla Tactics.** The success of the blitzkrieg was because of the application of a combination of conventional weapon systems for heavy fire and guerrilla fighters to surround, isolate and cut off escape routes. With a small strength of motivated fighters, the ISIS could defeat brigade and division size Iraqi forces. The ISIS captured Mosul with a strength of just 600-800 fighters.[56] Where required, it even did not abstain from using chemical weapons. It exploited sympathy from the erstwhile Bath party cadre and local supporters to create an extremely credible intelligence network.
- **Exploiting the Weak Political Structure.** Post withdrawal of the coalition forces from Iraq, the prevailing political situation was very weak and fragments of the erstwhile Bath party were still active and creating problems for the new government. The sectarian violence was at its peak. Taking advantage of the situation, the ISIS amalgamated the Bath party leaders and cadres and also the military equipment of the erstwhile Iraqi army.[57]
- **Religious Warfare.** The military success was followed up by imposing the Sharia law on the captured territory and declaration of the Caliphate. The fight was branded as a jihad to gain recognition and religious legitimacy.[58] The ISIS used the call of religion to attract youth to join the group.

- **Psychological Warfare.** The ISIS used a mix of terror, religion and propaganda to create a psychological impact which helped it gain success in tactical battles.[59] It used cruel acts of terror which were beyond human imagination to create a psychological impact which eventually subdued, rather finished, any local resistance. The brutal acts of beheading and shooting at point-blank range were circulated through social media. In addition to spreading terror, it used social media very effectively to spread its propaganda internationally. The effectiveness of the propaganda can be gauged by the huge number of international volunteers who lined up to join it. The organisation was successful in creating romanticism amongst the youth to join it to fight a global jihad.
- **Disregard for International Law and Norms.** The ISIS showed utter disregard to any kind of international norms or laws. The war in Iraq and Syria, in any case, did not get covered under the definition of a war as per the Geneva Convention. Hence, there was no question of the ISIS following the norms. It committed genocide against the Yazidi community. It even used children as soldiers to fight Jihad.[60]
- **Information Warfare.** The ISIS probably adapted the lessons of the Lebanon 2006 war, wherein the Hezbollah exploited the information domain to out manoeuvre Israeli defence forces. The importance given to information warfare by the Al Qaeda and the ISIS is evident in the statement made by Ayman Al Zawahri.

"*We are in a battle, and more than half of this battle is taking place in the battlefield of the media... we are in a media battle for the hearts and minds of our umma*"[61]

—Ayman Al Zawahri in his letter to Al-Zarqawi in 2005

- A report published by Brookings brings out the themes used by the ISIS in their propaganda. The propaganda was focused and complex, the themes depict a logic and has remained contemporary with the story of success and failure of the ISIS story.[62] It, in the initial phase, created an urgency to attend to the cause of suffering being faced by the Sunni brotherhood, the videos based on ill treatment to Sunnis by Shiite officials was shown. Thereafter, brand ISIS was created which depicted the ISIS as the agency which can fight for the noble cause, the videos depicting the religious fundamentalism of the ISIS, their brutal treatment to non-believers, and implementation of Sharia and declaration of Caliphate were released. The victories achieved by the ISIS in the early period of 2014 were shown with motivating songs. One particular video showed a high definition video of live operations on the Bajii oil refinery shot by a UAV and this created a feeling of ecstasy amongst the young followers of the ISIS.[63] It also created romanticism amongst the Sunni youth across the world to join the ISIS. The ISIS also showed the flexibility to adapt to the newer version of information technologies, the use of new generation social media platforms like WhatsApp, Telegram Chat and Twitter which were exploited to the optimum effect. The Internet was also used as a weapon when ISIS followers were able to hack into the website of the US Central Command (CENTCOM).[64] The thematic representation of the theme was highlighted in the Brookings report on ISIS propaganda.

Figure 2.3: Showing the themes followed by the ISIS in its propaganda.

Urgency (To create an urgency for a cause / build up narrative)	•To save Sunnis from atrocities being carried out against them •Sunni children being killed •To create a hatred against Shiite community
Agency	•Call upon individual Muslims to be the agency to protect Sunni Muslims •Project ISIS as the agency where they can join for the holy cause
Authenticity (to build the brand ISIS as the agency which will address the cause of Sunni)	•To build trust of Muslims with reference to the ISIS •The implementation of Sharia and declaration of Caliphate to prove their integrity towards Sunnis •Projecting extreme violence on non-believers was one such way
Victory	•Project the victories at Fallujah, Mosul, Raffa, Sinjhar with lot of pomp and show •Powerful videos showing the fighters, weapons, equipment with motivating songs, projecting ISIS as the divinely sanctioned agency

Source: The table has been culled out from the Brookings Report on Propaganda by ISIS[65]

- **Quest for Resources: Organised Criminal Activity.** It was reported that in 2014, the ISIS made more than $ 2 billion from various acts of criminal activity. The money primarily came from oilfields, looting of banks, extortion and forced taxation from the population residing in the captured territory.[66] It created a complex taxation system to cover virtually everything from commercial activity to education of children.[67] The ISIS also traded in valuable antiques, which were looted from the cities and archaeological sites they had captured.[68]

The ISIS, since its inception, has shown a character which was different from other transnational terrorist organisations. The brand image that it had created within such a short span is indeed a case study for informational warfare. It evolved in 2005, when it was primarily an anti-Shi-ite-group; it organised itself into a transnational terror group in 2013 and by 2014 it was an organisation which was running the captured territory like a government, managing the various functions of a state. It has shown tremendous flexibility and adaptability in keeping abreast with evolving technology. It uses drones to capture videos of raids by ISIS fighters for propaganda and use of the latest social media platforms to create a brand image to entice vulnerable young audiences captive to the virtual world. The recruitment saw volunteers from South Asia in the East to Canada in the West. They were also successful in creating a transnational network of terror, wherein small terror groups got connected and carried out lone wolf attacks under the ISIS banner. The ISIS may have been defeated for the time being; however their signature and footprint are likely to remain for long time. Its resurgence in other vulnerable regions like South East Asia may not be ruled out.

Analysis of Transformation in Warfare Post Cold War Period

In this chapter, the case studies of some of major conflicts post the end of the Cold War have been discussed and the major characteristics of warfare have been discussed. It is evident that the spectrum of means employed in these conflicts is much more varied then conventional warfare. The duration of the conflicts are also prolonged and the conflict was not always between two nation states. The adversaries now involved terrorist organisations like the ISIS, Hezbollah or Liberation Tigers of Tamil Elam (LTTE). There are states which give full-time support to terrorist organisations or use proxies against rival states to achieve their strategic national aims. Such transformation in warfare has been a subject of study amongst all the major powers of the world. This transformation of warfare has been described by Frank Hoffman in his seminal work of 2007, "*Conflict in 21st Century: Rise of Hybrid Warfare*". He defines hybrid warfare as:

> *"Hybrid threat incorporates a full range of different modes of warfare including conventional capabilities, irregular tactics and formations, terrorist acts including indiscriminate violence and coercion and criminal disorder. Hybrid Wars can be conducted by both states and a variety of non-state actors."*
>
> **—Frank Hoffman**

As discussed in chapter 1, there is a counter view, as some experts prefer not to use new adjectives and justify oversimplified depictions of warfare in two distinct bins: conventional and irregular. They say that there is no novelty in the concept of hybrid, as from time immemorial irregulars have been used with regular forces for speeding of operations or to cut off escape routes. In this regard, Hoffman argues that the use of irregulars with conventional forces fits in the concept of compound war as discussed by Thomas Huber in his book "*Compound Wars: That Fatal Knot*."[69] The 1971, India-Pakistan War or War of Liberation of Bangladesh, in which Bangladeshi freedom fighters organised as 'Mukti Bahini' fought as irregulars along with Indian Army, would qualify more as a compound war than a hybrid war, as non-conventional means like indiscriminate violence or coercion or criminal disorder were not employed by the Indian Army. There is another viewpoint which defines that hybrid war is a contemporary form of guerrilla warfare that employs both modern technology and modern mobilisation methods. The difference in the terminologies is very narrow and lies in the zone of ambiguity. Case studies bring out a pattern, which shows higher employment of non-kinetic means like psychological, cyber, propaganda and criminal activities to achieve political aims. Hoffman, in 2009, refined his definition of hybrid warfare as "*Any adversary that simultaneously and adaptively employs a fused mix of conventional weapons, irregular tactics, terrorism and criminal behaviour in the battle space to obtain their political objectives.*"[70] This definition brings out four mutually exclusive conditions which high-light the changing character of warfare. These is simultaneity, addictiveness, mix of various means, and criminal behaviour. These four aspects of the character of hybrid warfare is explained thematically in Figure 2.4.

Hoffman in his paper brings out that the participants or the adversary could be a state, a non-state actor, a supra-national actor or even a terrorist, and questions that showed the distinction based on the variation in employing the means of warfare or the hybridity in the nature of participants is more pertinent. The other aspects like simultaneity in employing a fusion or mix of various instruments of national power, being adaptive to employ either one or more than one or all four modes depending on the environment and situation, and the imposition of criminal behaviour to achieve the national aims are important factors to understand the evolving threat of hybrid warfare.

In this chapter, to carry out an analysis of the transformation of warfare post-cold war era, the definition of Hoffman as given in 2009, is being used to cull out various parameters required to analyse the conflicts. The case studies on Pakistan's

Figure 2.4: Character of Hybrid Warfare

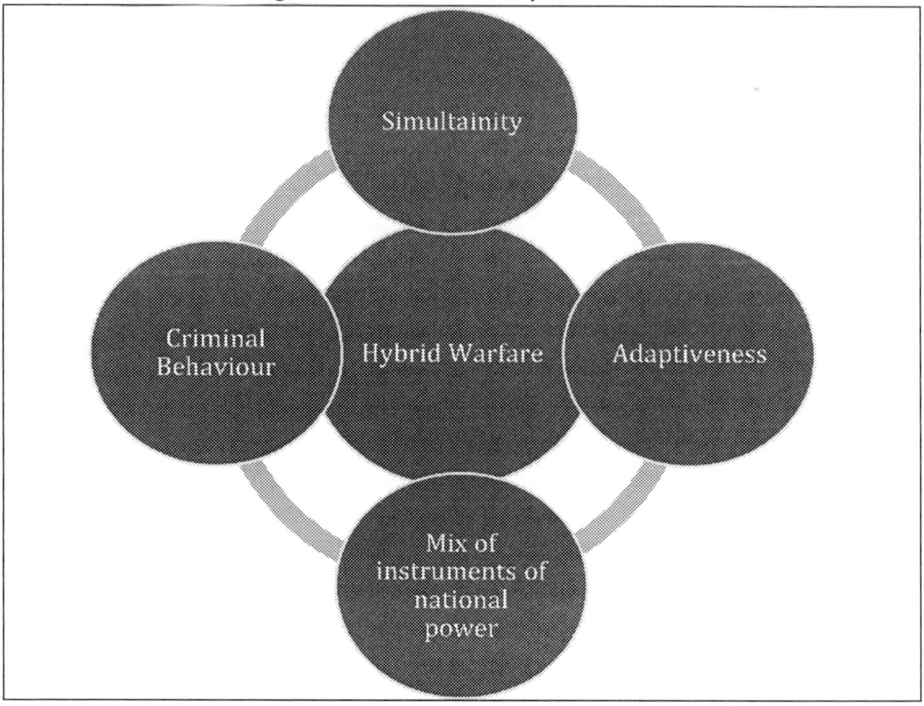

Source: Frank Hoffman in *Hybrid vs. Compound War*[71]

hybrid warfare with India and Afghanistan and China's hybrid strategy have been covered in separate chapters. However, the deductions from those chapters have been taken to analyse the issue holistically. It is evident that the various means/ instruments of warfare used in the major conflicts since end of the cold war have been more-and-more technical and non-kinetic. A snapshot of transformation that has emerged since post-cold war is summarised below:

- The argument given by Martin Van Creveld that with the advent of nuclear weapons, wars will reduce or may come to an end, has not happened. The number of conflicts since 1945 indicate that while wars between nuclear powers and their allies may not have happened, the conflicts/war have been taking place. This shows that there is a limit to deterrence. Hence, interstate conflicts will continue.
- States are focusing more on achieving political aims rather than destruction of the enemy.
- States are no more solely depending on conventional/kinetic means, but are increasingly using sub-conventional forms and measures short of war

such as cyber, economic and political coercion, to achieve national aims. The states are using proxy forces (irregular or terrorist organisations) and measures short of war in combination.

- The Hezbollah, Taliban, ISIS and Al Qaeda have shown that wars may not be limited to interstate, but the frequency of non-state transnational organisations challenging states is increasing.
- Irregulars, terrorists and non-state transnational organisations are gaining powers similar to states. The strategy and tactics used by these organisations are maturing and evolving with time.
- The strength of irregular, terrorist and non-state organisations are unconventional and irregular means. They have used information, cyber and economic warfare innovatively and adaptively.

To understand the transformation in warfare across the spectrum of conflict, it is imperative to understand the objective, the adversary and his concept of victory. To simplify the complexity of nature of conflict based on the intensity, violence and kinetic means, a model given by Dr. Hoffman is given below (Figure 2.5).

Figure 2.5: Heuristic Model of Continuum of Conflict

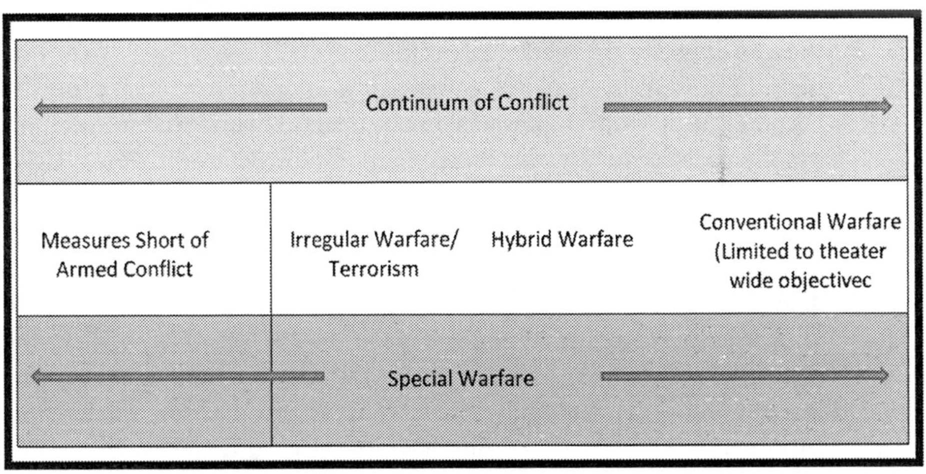

Source: Dr. Frank Hoffman in *Conflict Gray Zone and Hybrid Challenges*[72]

The continuum given above in not a rigid one, there cannot be a division between the four forms of warfare given in it. But this continuum given by Dr. Frank Hoffman is helpful in a way that it helps us understand the different levels of violence, starting from measures short of war like political, economic or cyber war to irregular, insurgency or acts of terrorism to full-fledged conventional wars.[73]

The analysis in this chapter brings out that most of the conflicts fall short of conventional war and are in the zone of hybrid warfare.

Emerging Character of Hybrid Warfare. Hybrid warfare covers the full spectrum of the warfare to include conventional capabilities, irregular tactics, terrorism and acts of criminal activities. Hybrid wars combine the vigour and lethality of conventional warfare with the ingenuity and fanatical fervour of irregular warfare.[74] The analysis of case studies in this chapter brings out higher employment of non-kinetic means like psychological, economic, political, cyber, propaganda and criminal activities to achieve political aims. Some of these trends which are being used aggressively by China, Iran, Russia and Pakistan along-with transnational terrorist groups are given in subsequent paragraphs:

> *Cyber warfare involves the actions by a nation-state or international organization to attack and attempt to damage another nation's computers or information networks through, for example, computer viruses or denial-of-service attacks.*
>
> **—Rand Corporation Report**

- **Use of Cyber Space and Information Technology.** Israeli's targeting the centrifuges at the Natanz plant in Iran in 2010 was possibly the first instance of the cyber tool being used as a weapon.[75] Since then, cyber has become one of the most powerful instruments of hybrid warfare. A report on 'Cyber Warfare' by the Rand Corporation defines it thus: "*Cyber warfare involves the actions by a nation-state or international organization to attack and attempt to damage another nation's computers or information networks through, for example, computer viruses or denial-of-service attacks.*"[76] The report by Rand brings out that the United States of America indicted 13 Russians for attempting to interfere in the 2016 election.[77] Russia has demonstrated the capability of cyber operations in forcing its will during the operations in Georgia, and later in Crimea. China has created Information Warfare (IW) militia units since 2002. The Chinese IW militia units are tasked with offensive and defensive CNO and EW responsibilities and psychological warfare and deception operations.[78] As the overall dependence on mobile networks and computers is increasing every day, so is the power of cyber warfare. With 5G, Internet of Things (IoT) and Artificial Intelligence infiltrating into the system, the vulnerability against cyber warfare will further increase.
- **Geo-economics.** Geo-economics is defined as the use of economic

instruments to promote and defend national interests, and to produce beneficial geopolitical results and the effects of other nation's economic actions on a country's geopolitical goals.[79] The use of geo-economics can be traced back to the Industrial Revolution, when the industrialised nations in Europe were competing over establishing their colonies or markets worldwide. The Dutch dominated commercial marine activity in the sixteenth and seventeenth centuries. England too promulgated the Navigation Act of 1651 and later the Staple Act of 1663 which restricted, prohibited and made commercial activities unviable for non-British and non-French companies. Navigation policies by France and England were directed primarily against the Dutch.[80] This was, possibly, the first recorded incident of economic warfare. The dominance of the Dutch was lost without a fight. Since then, economic warfare in forms of sanctions, ban on trade of critical items, imposing of tariff or artificially moderating the value of currencies to promote their own national interest have been used time and again by various powers. In recent times, China has been blatantly using its new-found economic status to coerce other nations. It has been slowly squeezing the international space available to Taiwan by its economic encirclement strategy.[81] China also used economic coercion on Japan by banning the export of rare earth minerals during the Senkaku Islands confrontation and against South Korea when THAD missiles were deployed.[82]

- **Political Manoeuvre.** A Rand report on political warfare defines it as follows: "Political warfare consists of the intentional use of one or more of the implements of power-diplomatic/political, information/cyber, military/intelligence, and economic to affect the political composition or decision making in a state."[83] (Figure 2.6). In recent times, Iran, Pakistan and Russia are amongst the few nations which have been carrying out political warfare to influence not only the decision making in a state but to get a government of their preference in the target state. Iran has been interfering in Syria and Iraq through its proxies through combination of socio-economic, religious and cultural influence.[84] The Russian annexation of Crimea, efforts towards a pro-Russian government in Ukraine and assistance to South Ossetia and Abkhazia to declare independence have seen the employment of cyber, proxies, propaganda and covert use of non-uniformed soldiers.[85]

Figure 2.6: Political Warfare: Thematic Representation

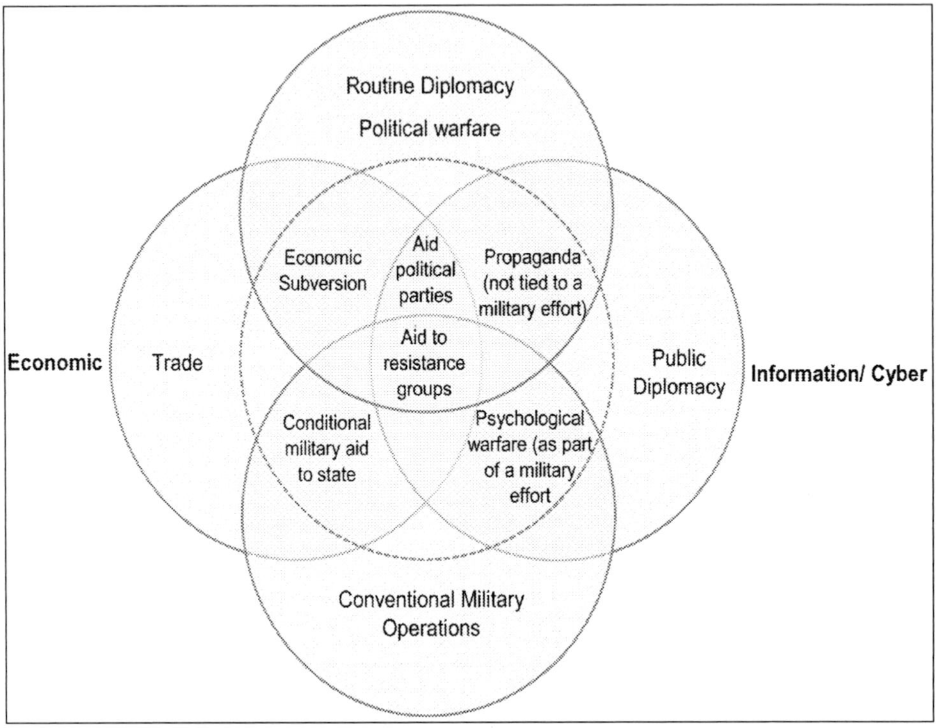

Source: Rand Corporation report: On Modern Political Warfare[86]

- **Psychological/Propaganda.** This is one of the most important facets of hybrid warfare and has been part of almost all recent conflicts. The target is the psychological or the cognitive domain. The information and cyber operations converge to project a 'just cause' to give legitimacy to the campaign. The propaganda carried out by the ISIS by using electronic and social media to create its brand image and to bring legitimacy is an apt case study.[87] The ISIS was hugely successful in generating a kind of romanticism to its cause, which drew thousands of volunteers across the globe to join which included highly educated professionals from Western countries. Pakistan has also been involved in psychological warfare and propaganda in Jammu and Kashmir.[88]
- **Legal Warfare.** China in its Defence White Paper 2006 promulgated the 'three warfare strategy', which along with 'Psychological' and 'Media' highlights the use of Legal warfare to enlarge the strategic concept of 'People's war'. Legal warfare aims to exploit the legal system to achieve

political or commercial objectives. China has been a strong proponent of Legal warfare and has used it in South and East China Sea disputes.

Impact of Emerging/Disruptive Technologies. The term "*disruptive technology*" was coined by the Clayton M. Christensen in his famous book "*The Innovator's Dilemma*" and refers to any enhanced or completely new technology that replaces and disrupts an existing technology, rendering it obsolete. It is designed to succeed similar technology that is already in use. The emerging disruptive and cutting-edge technologies are finding their application in warfare. The USA, Russia, China and many other nations are investing in a big way to adapt these technologies into military hardware. These technologies will change the future battle field, which will lean more towards non-contact warfare than contact warfare. Some such disruptive technologies and their impact on the future of hybrid warfare are discussed below.

- **Artificial Intelligence.** Artificial intelligence (AI) is the simulation of human intelligence processes by machines, especially computer systems. It has come a long way and when adapted into military hardware, can produce a multiple effect. The combination of AI with big data analysis can increase the accuracy, efficiency and reliability of the weapons systems. AI has a tremendous application in increasing battle field transparency and reducing the OODA (observe, orient, decide and action) loop. This will improve the decision making capability of commanders during the fog of war manifold. In the days ahead, innovative use of AI and its adaptation will be the key in retaining the initiative during the conduct of hybrid war by both the state and non-state actors.
- **Quantum Technology.** Quantum technology will transform the computing and communication. Quantum computing will create processors which will be significantly faster and can hold an enormous amount of information. It will be a game changer in cryptography as it will create complex algorithms for advanced encryption. The quantum communication technology will create un-hackable communication networks. China and the USA are investing heavily in quantum technologies.
- **Autonomous Weapons.** The autonomous technology has already revolutionised intelligence, surveillance, reconnaissance and remote targeting. Unmanned aerial vehicles and unmanned submersible vehicles

are already in use. China has an ambitious underwater Manned/ Unmanned Submersibles program. There are confirmed reports of China using unmanned underwater vehicles in the Indian Ocean.[89] Armed autonomous drones have emerged as force multipliers in Iraq, Syria and Afghanistan. This being a cheap technology, it is being used by terrorist organisations. Autonomous drones loaded with explosives have emerged as a serious threat to security forces.

- **Swarming Technology**. Swarming by drones is explained 'as multiple unmanned platforms and/or weapons deployed to accomplish a shared objective, with the platforms and/or weapons autonomously altering their behaviour based on communication with one another.'[90] The access to communication facilitates the swarm of drones to concentrate on a target while taking off from different places. The technology has immense potential as it is cheap and versatile. Military analysts also envisaged the use of micro drone swarms for biological and chemical warfare by terrorist organisations and rogue states.
- **Hypersonic Strike Technology**. A hypersonic refers to speeds of Mach 5 and higher—five times faster than the speed of sound (3,836 mph), which is around 1 mile per second. A new weapons race has started between the USA and China for missiles which can defeat anti-ballistic missile systems. While the weapon is being envisaged for a conventional war, however the technology once developed can be used for hybrid war also. This technology can be used for political signalling by carrying out precision targeting of military facilities thereby reducing collateral damage
- **Direct Energy Weapons**. High energy laser weapon systems are being quoted as the newest and most promising innovations in the armaments and technology field. High energy laser weapons offer the potential of low cost, speed of light with multi shot option thereby increasing the accuracy and precision engagement of strategic and tactical targets. These are capable of being mounted on a variety of air, land or sea-based platforms. In an irregular war or hybrid war scenario, these laser weapons can be used very effectively against UAVs.
- **Drone Technology**. Drone technology is not new, however with the quantum jump that this technology has made, it is proving to be one of the game changers in hybrid war or for Non-Contact Warfare. The year 2019 saw innovative and highly effective use of drones, which has given

a new dimension to threats from air, which is more difficult to detect, identify and respond to. The precise and targeted strike by drones by Houthi terrorists on the Aramco oil facility of Saudi Arabia on Sep 14 and earlier in January, during a parade killing six Yemeni soldiers show the potency of drone technology. Another innovative use was seen close to the India-Pakistan border in the state of Punjab where drones were used by terrorist groups to ferry weapons into India from Pakistan. The high tech fence and Air Force EW radars were rendered ineffective to detect, identify and respond to the intrusion by drones. In recent reports in the *South China Morning Post*, a major attraction of China's military parade of October 1, was the display of supersonic and stealth drone technology.

Hybrid Warfare in Indian Context. The concept of hybrid warfare in not new to India. As discussed in Chapter 1, Kautilya, the master strategist was the strongest proponent of hybrid war. Of the four types of warfare described by him in the *Arthashastra*, the first three fall in the domain of modern-day hybrid warfare and conventional war came as the last option. Following the advice of Kautilya, Emperor Chandragupta Maurya, in the third century BC, created one of the largest empires in the history of the world. It remains fathomless why Kautilya was not as revered in India as Sun Tzu in China or Machiavelli or Clausewitz in Europe or the West. Arjun Subramanian, *in India's Wars: 1947 to 1971*, mentions that Pandit Nehru was deeply influenced by Gandhian principles like Satyagraha, non-violence and non-cooperation based on which India fought its war of independence.[91] Hence, to Nehru, the *Arthashastra's* ruthless wielding of power and overwhelming dominance of state over welfare of the people did not appeal.[92] As a result, in independent India, 'War' or 'Power' as tools of statecraft remained on the periphery. It was only in 1971 that Indira Gandhi looked at hard power as a tool of statecraft.

India's Experiment with Hybrid Warfare: Retrospective Analysis of War of Independence of Bangladesh. The National Military Objectives (NMO) as given in the Joint Doctrine of the Indian Armed Forces, articulates that the primary role of the Indian Armed Forces is to defend the territorial integrity and deter war by building suitable deterrence across the complete spectrum of war.[93] According to the Doctrine, the traditional roles of India's military remain confined to defence of the borders, combating internal disorder and providing aid to civil authorities in times of distress, when called upon by the government. The organisation, structure and capability building of the Indian Armed Forces are focused and templated to conventional threat and 'state-on-state warfare',[94] All the offensive operations that have been launched by India since independence have been reactive.[95] Maj. Gen. Alok Deb in his article on 'India and Hybrid Warfare' has said that, "*a common criticism about the Indian Army over the years has been its propensity to rigidly adhere to a traditional mindset.*" It was only in 1971 in the war with Pakistan in which India carried out the prior preparation of the battle field on the Eastern Front. However, the offensive was launched after the IB was first crossed by Pakistan on the Western Front. The War of Independence of Bangladesh in 1971, in retrospective analysis, emerges to be an apt case study of hybrid war.

Erstwhile East Pakistan (Bangladesh) suffered one of the worst humanitarian crises of the 20th century.[96] In the general election of 1970, the Awami League led by Sheikh Mujibur Rehman, achieved a clear majority which was intolerable for and unacceptable to the ruling class of West Pakistan. Finding no other option, Gen. Yahya Khan declared martial law in East Pakistan on March 23, 1971. The worst ever genocide of South Asia commenced with the killing of Bengali intellectuals and academics of Dacca University. This triggered a mass migration of Bengali refugees to India. By November 1971 more than 10 million refugees had taken shelter in India.[97] Having tried all diplomatic options of trying to resolve the problem in East Pakistan, Prime Minister Indira Gandhi was left with no choice but to solve the problem militarily. The major actions taken by the Government of India led by Mrs. Indira Gandhi in 1971, in preparation of the battle field and subsequently liberating Bangladesh are discussed below with reference to the tenets of hybrid warfare.

- **Proactive Diplomacy.** Prime Minister Mrs. Indira Gandhi along with her Foreign Minister visited all the major countries to highlight the atrocities being carried out by Pakistan and tried to build a consensus to bring a motion in the UN against the actions of Pakistan.[98] At that particular point of time, the diplomatic calculus was in favour of Pakistan. The increasing closeness between the USA and China was in favour of the China-Pakistan nexus. This emerging relationship between the USA and China completed the triangular relationship of USA, China and Pakistan.[99] Mrs. Gandhi being a realist could foresee the consequences of this and the possibility of finding India cornered in case of another Indo–akistan war. She took the timely and apt decision to sign the historic

Indo–Russian Friendship Treaty in August 1971.[100] This not only provided the required assurance of a veto in the UN but also opened the route for the much required military hardware for the Indian Armed Forces.[101] Mrs. Indira Gandhi with her astute real politic approach was able to outmanoeuvre the Pakistani diplomatic advantage. Having catered for Russian support, she focused her attention towards East Pakistan.

- **Empowering Mukti Bahini.** Mukti Bahini is a Bengali word, the English translation of which is 'Freedom Fighters'. Mukti Bahini comprised both civilian as well as regular Pakistani army personnel who had defected from the Pakistani Army and joined the war of independence.[102] These officers and soldiers were mostly from the East Bengal Rifles (EBR) which had mutinied after the crackdown started in East Bengal. By mid-April 1971, the Mukti Bahini reorganised itself as the Bangladesh Armed Forces, however, Mukti Bahini remained a more popular name for this force.[103] The role of the Research and Analysis Wing (RAW) in the initial stages and the Indian Army close to D day with respect to training, equipping and providing logistical support is well documented.[104] The Mukti Bahini was a very effective guerrilla force which frequently carried out raids and ambushes and continuously harassed the Pakistani forces. One of the major contributions by it was in terms of being the eyes and ears of the Indian Army. Cadres of the Mukti Bahini moved ahead of the advancing columns and provided immediate tactical level intelligence, which acted as a force multiplier. It assisted the Indian Army to manoeuvre beyond the strongly-held positions thus avoiding attrition battles and ensuring early capitulation of the strongholds.[105] The air and naval wings also contributed to the army component of the Mukti Bahini. The Mukti Bahini frogmen comprised ex-submariners and expert divers who mined the riverine areas of Chittagong, Chalna and Khulna and successfully interdicted the supplies from sea for the Pakistani forces operating in the Eastern part. The frogmen were successful in destroying approximately one hundred thousand tons of supplies bringing shipping activities virtually to a halt. This naval blockade restricted the reinforcement only to air.[106] The naval element of the Mukti Bahini later joined the Indian Navy during the course of war.[107] Those Bengali men who defected from the Pakistani Air Force and airlines were organised as Kilo Force and located at an airstrip near Dimapur. The Indian Air Force assisted them with one Dakota, an Otter light transport aircraft and an Alouette helicopter. This small component assisted and fought the war along with the Indian Air Force.[108] The optimum use of the Mukti Bahini as an irregular force along with the conventional force, facilitated the strategic aim of India to attain a decisive victory over Pakistan in just 14 days that led to creation of a democratic nation.

Hybrid Threat to India. India has fought four wars (including Kargil in 1999) and one war with China since independence. The Line of Control (LC) with Pakistan continues to remain active with incidents like cross-border firing, artillery shelling, raids on border posts, infiltration and other such activities which can be graded activities short of war or as a situation of No War and No Peace (NWNP). Pakistan continues to wage a proxy war in Jammu and Kashmir and terrorist groups sponsored and sheltered by Pakistan have been indicted for terrorist acts in India. On the Line of Actual Control (LAC) along with Actual Ground Line of Patrolling (AGPL) with China, the incidents of face-offs between the patrolling teams of both nations are frequent; some of the face-offs have even resulted in the physical altercations. However, armed conflict has not taken place, except during the 1962 war and the Nathu La incident of 1967. Chinese covert support to insurgent groups of Nagaland, Manipur, Mizoram, Assam and the Naxal movement are well documented. In recent times, China has also been involved in offensive cyber operations against India. A detailed analysis of Pakistan's tryst with hybrid warfare and Chinese characteristics of hybrid warfare have been covered in detail along with the likely threat they pose to India, in chapters 3 and 4, respectively.

NOTES

1. Carl von Clausewitz, On War, ed. and trans. Michel Howard and Peter Paret, Princeton, 1976, p 89
2. Hew Strachan in *The Changing Character of War*, EUROPAEUM lecture, delivered at The Graduate Institute of International Relations, Geneva, Nov. 9, 2006
3. For more details, see Hew Strachan, in *The Changing Character of War*, EUROPAEUM lecture, delivered at The Graduate Institute of International Relations, Geneva, Nov. 9, 2006
4. Martin Van Creveld in *Transformation of War*, The Free Press, 1991, p. 8
5. Ibid
6. Ibid
7. For more, see Martin van Creveld, On Transformation of War, The Free Press, 1991
8. Brig Gurmeet Kanwal, in *Hybrid Warfare The Changing Character of Conflict*, Pentagon Press,

p. 4

9. Mary Kaldor, in review essay, *In Defence of New Wars*, Published in *Stabilit*, 2(1): 4, pp.1-16, available at http://dx.doi.org/10.5334/sta.at accessed on 4 January 2019
10. Brig. Gurmeet Kanwal, in *Hybrid Warfare The Changing Character of Conflict*, Pentagon Press.
11. Brig. Gurmeet Kanwal, in *Hybrid Warfare The Changing Character of Conflict*, Pentagon Press, p. 4
12. Kalb, M. & Saivetz, C. (2007). The Israeli—Hezbollah War of 2006: The Media as a Weapon in Asymmetrical Conflict. *Harvard International Journal of Press/Politics*, 12(3), 43–66. Accessed on 21 June 2019.
13. For a detailed account of the Israel-Hezbollah war, see Andrew Exum, *Hezbollah at War: A Military Assessment*, The Washington Institute for Near East Policy, available at https://www.washingtoninstitute.org/uploads/Documents/pubs/PolicyFocus63.pdf accessed on Feb, 10, 2019
14. S. Samuel & C. Rajiv, *Israel and the Challenges of Hybrid Warfare, Hybrid Warfare; The changing character of conflict*, Pentagon Press, p. 129
15. For a detailed account of the Israel-Hezbollah war, see Andrew Exum, *Hezbollah at War: A Military Assessment*, The Washington Institute for Near East Policy, available at https://www.washingtoninstitute.org/uploads/Documents/pubs/PolicyFocus63.pdf accessed on Feb. 10, 2019
16. Ibid
17. Ibid
18. Ibid
19. Kalb, M. & Saivetz, C. (2007). The Israeli—Hezbollah War of 2006: The Media as a Weapon in Asymmetrical Conflict. *Harvard International Journal of Press/Politics*, 12(3), 43–66.
20 Kalb M. & Saivetz, C. (2007). The Israeli—Hezbollah War of 2006: The Media as a Weapon in Asymmetrical Conflict. *Harvard International Journal of Press/Politics*, 12(3), 43–66.
21. Ibid
22. Ibid
23. Ibid
24. Ibid
25. Ibid
26. Ibid
27. Ibid
28. Major Gogoi of Rashtriya Rifles saved a group of officials of the Election Commission of India, who were conducting Parliamentary elections in Kashmir on April 7, 2017, from a mob throwing stones. He used one of the stone pelters as a shield so that the stone pelting stopped. For more details, see, "Outrage over Omar's Abdullah's video of army using a human shield", *National Herald*, Apr 14, 2017 available at https://www.nationalheraldindia.com/news/outrage-over-former-jk-chief-minister-omar-abdullah-video-of-army-using-a-human-shield accessed on 20 June 2019
29. Kalb, M. & Saivetz, C. (2007). The Israeli-Hezbollah War of 2006: The Media as a Weapon in Asymmetrical Conflict. *Harvard International Journal of Press/Politics*, 12(3), p. 62
30. For a detailed account of the Israel-Hezbollah war, see Andrew Exum, *Hezbollah at War: A Military Assessment*, The Washington Institute for Near East Policy, available at https://www.washingtoninstitute.org/uploads/Documents/pubs/PolicyFocus63.pdf accessed on Feb

10, 2019

31. Rain Ottis, Analysis of the 2007 Cyber Attacks Against Estonia from the Information Warfare Perspective, available at https://ccdcoe.org/.../Ottis2008_Analysis Of 2007, from The Information Warfare Perspective, Accessed on 21 June 2019
32. Rain Ottis, in Analysis of the 2007 Cyber Attacks Against Estonia from the Information Warfare Perspective, Cooperative Cyber Defence Centre of Excellence, Tallinn, Estonia available at https://ccdcoe.org/.../Ottis2008_Analysis Of2007 From The Information Warfare Perspective
33. Ibid
34. Aman Saberwal, in *Russia and Hybrid Warfare: Achieving Strategic Goals without Fighting, Hybrid Warfare: Changing Character of Conflict*, Pentagon Press, 2018, p. 67
35. For more details on the geography and history of South Ossetia see South Ossetia, Geo History, available at https://geohistory.today/south-ossetia/ accessed on June 21, 2019
36. Ibid
37. For more details on the background and history of the conflict, see The Russo-Georgian War (2008): The role of the cyber-attacks in the conflict, AFCEA, available at https://www.afcea.org/committees/cyber/documents/therusso-georgianwar2008.pdf accessed on June 20 2019
38. For more details on the Georgia-Russia 2008 war, see Russia-Georgia conflict of 2008: Context and Implications for US, Congressional Research Service, available at https://fas.org/sgp/crs/row/RL34618.pdf accessed on June 21 2019
39. Aman Saberwal, in *Russia and Hybrid Warfare: Achieving Strategic Goals without Fighting, Hybrid Warfare: Changing Character of Conflict,* Pentagon Press, 2018, p. 67
40. Ibid
41. For more details on the Georgia-Russia 2008 war, see Russia-Georgia conflict of 2008: Context and Implications for US, Congressional Research Service, available at https://fas.org/sgp/crs/row/RL34618.pdf accessed on June 21 2019
42. The data has been taken from site of Crimean Historical Society, available at http://www.crimeahistory.org/population-of-crimea/ accessed on 20 Jun 2019
43. For more details on Crimea's special place in strategic calculus of Russia, see Sergey Saluschev, Annexation of Crimea: Causes, Analysis and Global Implications.' *Global Society Journal,* Vol. 2, 2014, available at https://escholarship.org/uc/item/5vb3n9tc, accessed on June 19, 2019
44. Daniel Treisman, in Why Putin Took Crimea, Foreign Affairs, May-Jun issue, available at https://www.foreignaffairs.com/articles/ukraine/2016-04-18/why-russian-president-putin-took-crimea-from-ukraine accessed on June 19, 2019
45. For more details on Crimea's special place in strategic calculus of Russia, see Sergey Saluschev, Annexation of Crimea: Causes, Analysis and Global Implications'. *Global Society Journal,* Vol. 2, 2014, available at https://escholarship.org/uc/item/5vb3n9tc, accessed on Jun 19, 2019
46. Ibid
47. Aman Saberwal, in *Russia and Hybrid Warfare: Achieving Strategic Goals without Fighting, Hybrid Warfare: Changing Character of Conflict*, Pentagon Press, 2018, p.65
48. Night Wolves Motorcycle Club is based in Russia. They have received substantial funding from the Kremlin with estimates ranging from tens of millions of roubles to several hundred million roubles. The group is a staunchly Russian nationalistic organization. For more details on the Club and its role in Crimea, see, Joseph V. Micallef, in The Night Wolves: Russian Motorcycle Club or Kremlin Militia? Available at https://www.military.com/daily-news/2018/

09/04/night-wolves-russian-motorcycle-club-or-kremlin-militia.html accessed on 20 June 2019.
49. For more details on Cyber War in Crimea, see Crimea-Russian Cyber strategy to hit Ukraine, available at https://resources.infosecinstitute.com/crimea-russian-cyber-strategy-hit-ukraine/#gref accessed on 20 June 2019
50. Ibid
51. Aman Saberwal, in *Russia and Hybrid Warfare: Achieving Strategic Goals without Fighting, Hybrid Warfare: Changing Character of Conflict*, Pentagon Press, 2018, p. 65
52. Time Line: The Rise and Fall of Islamic State, *Reuters*, March 2, 2019, available at https://www.reuters.com/article/us-mideast-crisis-islamic-state-timeline/timeline-the-rise-and-fall-of-islamic-state-in-iraq-and-syria-idUSKCN1QJ08N accessed on 22 June 2019; History of ISIS, in 'Since /911, available at https://since911.com/explore/history-isis accessed on 22 June 2019
53. Ibid
54. Ralph D. Thiele, 'The New Colour of War-Hybrid Warfare and Partnership', ISPSW Strategy Series, issue 383, October 2015, available at https://www.flies.ethz.ch/isn/194330/383_Thiele.pdf, accessed on 22 June 2019
55. Ibid
56. Ahmed Salah Hashim in State and Non State Hybrid Warfare, Available at https://www.oxfordresearchgroup.org.uk/blog/state-and-non-state-hybrid-warfare accessed on 22 June 2019
57. Ralph D. Thiele, 'The New Colour of War: Hybrid Warfare and Partnership', ISPSW Strategy Series, issue no 383, October 2015, available at https://www.flies.ethz.ch/isn/194330/383_Thiele.pdf, accessed on 22 June 2019
58. Ibid
59. Ibid
60. Shruti Pandalai, in *The Mutating Wars, Hybrid Warfare: The Changing Nature of Conflict*, Pentagon Press p. 82
61. Translation of letter written by Ayman Al-Zawahri, "Letter from Al-Zawahiri to Al-Zarqawi," For translation of the complete letter, see *Global Security*, October 11, 2005, http://www.globalsecurity.org/security/library/report/2005/zawahiri-zarqawi-letter_9jul2005.htm. Accessed on 1 June 2019
62. For more details on the propaganda carried out by the ISIS, refer to Alberto M. Fernande, in 'Here to Stay and Growing: Combating ISIS Propaganda Networks' The Brookings Project on US Relations with Islamic World, US Islamic World Forum Papers, 2015 pp 11, available at https://www.brookings.edu/wp-content/uploads/2016/06/IS-Propaganda_Web_English.pdf accessed on 15 June 2019
63. Scott Jasper and Scott Mooreland, in ISIS: An Adaptive Hybrid Threat in Transition, Small War Journals, available at https://calhoun.nps.edu/bitstream/handle/10945/50642/Jasper_Moreland_ISIS_%20 An_Adaptive_Hybrid%20Threat_2016-10-29.pdf?sequence=1&isAllowed=yaccessed on 15 June 2019
64. Ibid
65. For more details on the propaganda carried out by the ISIS, please refer to Alberto M. Fernande, in 'Here to Stay and Growing: Combating ISIS Propaganda Networks' The Brookings Project on US Relations with Islamic World, US Islamic World Forum Papers, 2015 pp 11, available at https://www.brookings.edu/wp-content/uploads/2016/06/IS-Propaganda_Web_English.pdf accessed on 15 June 2019

66. Scott Jasper and Scott Mooreland in ISIS: An Adaptive Hybrid Threat in Transition, Small War Journals, available at https://calhoun.nps.edu/bitstream/handle/10945/50642/Jasper_Moreland_ISIS_%20An_Adaptive_Hybrid%20Threat_2016-10-29.pdf?sequence=1&isAllowed=y accessed on 15 June 2019
67. Ibid
68. Ibid
69. Frank Hoffman, in 'Hybrid vs. Compound War, *Armed Forces Journal.* Available at http://armedforcesjournal.com/hybrid-vs-compound-war/ accessed on 18 June 2019
70. Ibid
71. Ibid
72. Dr Frank Hoffman, in Examining Complex Forms of Conflict Gray Zone and Hybrid Challenges, Prism, Vol. 4, National Defence University, available at https://cco.ndu.edu/News/Article/1680696/examining-complex-forms-of-conflict-gray-zone-and-hybrid-challenges/ accessed on 2 Feb 2018
73. Ibid
74. Frank Hoffman, "Conflict in 21st Century: Rise of Hybrid Warfare", Potomac Institute for Policy Studies, p, 8, available at https://potomacinstitute.org/23-publications/reports/1267-conflict-in-the-21st-century-the-rise-of-hybrid-wars?jus1=2 accessed on 30 Oct 2018
75. BBC report on Suxtnet, Time line how Suxtnet attacked the nuclear plant, available at https://www.bbc.com/timelines/zc6fbk7 accessed on 20 June 2019
76. Getting Ready to Fight Cyber Warfare, Report by Rand Corporation, available at https://www.rand.org/topics/cyber-warfare.html accessed on 25 Nov 2018.
77 Ibid
78. Nicholas Lyall, China's Cyber Militias, The Diplomat, available at https://thediplomat.com/2018/03/chinas-cyber-militias/ accessed on 10 June 2019
79. Robert D. Blackwill and Jennifer M. Harris in *War by other Means: Geo-economics and Statecraft,* The Belknap Press of Harvard University Press, 2016 edition, p. 21
80. Laura La Haye on mercantilism, available at https://www.econlib.org/library/Enc/Mercantilism.html accessed on 18 June 2918
81. For more details see *Geoeconomics in Chinese Foreign Policy,* Robert D. Blackwill and Jennifer M. Harris in *War by other Means: Geo-economics and Statecraft,* The Belknap Press of Harvard University Press, 2016, p. 93
82. Shubhankar Basu, in Economic Warfare: The Most Effective Weapon in Chinese Strategy of Unrestricted warfare, https://www.vifindia.org/article/2019/march/04/economic-warfare-the-most-effective-weapon-in-chinese-strategy-of-unrestricted-war
    For more details, see *Geo-economics' in Chinese Foreign Policy,* Robert D. Blackwill and Jennifer M. Harris in *War by other Means: Geo-economics and Statecraft,* The Belknap Press of Harvard University Press, 2016, 93
83. Rand Corporation report on, The Growing Need to Focus on Modern Political Warfare, available at https://www.rand.org/pubs/research_briefs/RB10071.html accessed on 17 June 2019
84 Ibid
85. Ibid
86. Ibid
87. The case study of the ISIS has been discussed in this chapter, For more details on the propaganda carried out by the ISIS please refer to Alberto M. Fernande in 'Here to Stay and Growing:

Combating ISIS Propaganda Networks', The Brookings Project on US Relations with Islamic World, US Islamic World Forum Papers, 2015 pp 11, available at https://www.brookings.edu/wp-content/uploads/2016/06/IS-Propaganda_Web_English.pdf accessed on 15 June 2019

88. Hybrid warfare by Pakistan has been covered in detail in chapter 3.
89. Captain Kamlesh K. Agnihotri, in High Technology Developments in China: Leveraging or Military Effectiveness, Issue Brief, Centre for Joint Warfare Studies, Mar 2019, p. 14
90. Zachary Kallenborn and Philipp C. Bleek in, 'Drones of Mass Destruction: Drone Swarms and the Future of Nuclear, Chemical, and Biological Weapons in *War on the Rocks*, Feb 14, 2019, available at https://warontherocks.com/2019/02/drones-of-mass-destruction-drone-swarms-and-the-future-of-nuclear-chemical-and-biological-weapons/ accessed on June 29, 2019
91. Arjun Subramaniam, in *India's Wars: 1947 to 1971*, Harper Collins Publishers India, p. 442
92. Ibid
93. For all the NMOs as given in the Joint Doctrine of the Indian Armed Forces 2017, please see p. 13 of the Joint Doctrine available at https://bharatshakti.in/wp-content/uploads/2015/09/Joint_Doctrine_Indian_Armed_Forces.pdf accessed on Jan 22, 2019.
94. Maj. Gen. Alok Deb, in *India and Hybrid Warfare, Hybrid Warfare: The Changing Nature of Conflict*, Pentagon Press, p.173
95. For a detailed account of all wars fought by India since independence, see Arjun Subramaniam, *India's Wars: 1947 to 1971*, Harper Collins Publishers India
96. Ibid
97. Ibid
98. Ibid
99. Arun Mohanty in Toasting Legacy of 1971 Indo-Soviet Friendship Treaty available at https://www.rbth.com/articles/2011/08/09/toasting_legacy_of_1971_indo-soviet_friendship_treaty_12842 accessed on 26 June 2019
100. Ibid
101. Ibid
102. For a detailed account of action by the Mukti Bahini, see Arjun Subramaniam, *India's Wars: 1947 to 1971*, Harper Collins Publishers India, p. 355
177. Ibid
103. On the role of the Mukti Bahini in the India-Pakistan War 1971, see *Courage and Conviction* by Gen. V. K. Singh and *The India-Pakistan War of 1971* by S.N. Prasad and U.P. Thapliyal
104. Arjun Subramaniam in his book *India's Wars: 1947 to 1971*, Harper Collins Publishers, has quoted an interview with Lt. Gen. Shammi Mehta, Former Army Commander, which gives an excellent insight on the contribution of the Mukti Bahini
105. Garry J. Bass in, Blood Telegram: India's Secret War in East Pakistan, Random House India, pp 181-182
106. Arjun Subramaniam, India's Wars: 1947 to 1971, Harper Collins Publishers India, p. 357
107. For detailed account of action by Naval and Air Wing of Mukti Bahini, see, Arjun Subramaniam, India's Wars: 1947 to 1971, Harper Collins Publishers India, pp. 356-57

3

Pakistan: An Ardent Practitioner of Hybrid Warfare

"Pakistan has been created and its security and defence is now your responsibility. I want them to be the best soldiers in the world. So that no one can cast an evil eye on Pakistan, and if does, we shall fight him to the end, until he throws us into Arabian Sea or we drown him in Indian Ocean."[1]

—Governor General Muhammad Ali Jinnah
Founder of Pakistan

ABSTRACT

This chapter argues that the strategic culture of the Pakistan Army is the source of the turbulent relationship that Pakistan has with India and Afghanistan. The chapter also argues that Pakistan's behaviour is unlikely to change even if the Kashmir issue is resolved. The argument has been supported by the analysis of Pakistan's involvement in the internal affairs of Afghanistan and the turbulent relations with India since its creation. How the Pakistan Army has evolved the methods of hybrid war against Afghanistan and India has also been covered.

The above statement by the founder of the nation immediately after the formation of Pakistan shows the foundation or the belief of Pakistan towards India. As a nation it never accepted that it can ever co-exist with its neighbours peacefully. Since its formation Pakistan has had complex relations with both its neighbours, India and Afghanistan. The genesis of this relationship can be attributed to the legacy of colonial rule and its hurried exit. While exiting, the border based on the

Durand and Radcliffe lines by the British government has been one of the major reasons for the unstable Indian subcontinent. However, the way Pakistan has employed various methods of hybrid warfare to exploit these complex issues is a matter of study. This chapter would argue that Pakistan has successfully employed hybrid warfare to achieve its strategic aims. It has been able to interfere successfully in the internal affairs of Afghanistan and converting it into a battleground of the superpowers. Thus it has not only kept itself relevant, but has also gained the required financial and material support to initiate and support a terrorist movement in state of Jammu and Kashmir and other parts of India. In order to analyse the actions of Pakistan, it is necessary to understand the strategic culture of the Pakistan Army which is influenced by deep-rooted security perceptions. It is even more important in the case of Pakistan as it is the army that dominates the decision making with respect to domestic as well as foreign affairs in Pakistan.[2] As the Pakistan Army has controlled the state directly or indirectly for most of its history, it is well positioned to orchestrate the internal and external environment to pursue its own institutional ends.[3] Lt. Gen. (Retd.) Ata Hasnain of the Indian Army said, "*While other nations have armies, Pakistan army has a nation.*"[4] This explains the interference or dominance of the Pakistan army in everyday governance in Pakistan.

Strategic Culture of Pakistani Army

The concept of strategic culture is an important construct to explain the behaviour of states in international relations. It assists policy makers to understand how a state has behaved in the past and how it is likely to behave in the future.[5] Hasan Askari Rizvi in '*Pakistan's Strategic Culture in South Asia in 2020: Future Strategic Balances and Alliance*' has highlighted that Pakistan's strategic culture has been affected due to acute insecurity developed during the early nationhood, the wars with India in the post-partition period and the troubled relationship with Afghanistan. Peter R. Lavoy in '*Pakistan's Strategic Culture*', a report prepared for Defence Threat Reduction Agency, USA, has used theories of '*Neo Realism*' and '*Strategic Culture*' along with role of the elite, whom he calls '*mythmakers*', who redefine, modify and transform the behaviour of the state to their own strategic preferences and their understanding.[6] Ayesha Siddiqa in her book '*Military Inc: Inside Pakistan's Military Economy*' has highlighted the concept of '*Milbus*' and the role of military economy in the functioning of the state. While analysing Pakistan as the case study for her book, she has argued that Pakistan has a militaristic-totalitarian system, wherein the Army General is the head of the state.

In any such militaristic-totalitarian set-up, to preserve its relevance, the military elite will orchestrate or create myths which will shape the strategic culture of the state.[7] C. Christine Fair in her report to The National Bureau of Asian Research, Washington DC has argued that Pakistan is a revisionist state and it continues to secure control over Kashmir and also to resist India's stature in international forums.[8] Pakistan started wars in 1947-48, 1965 and then in 1999 against India in Kashmir and failed terribly. Having failed in all its conventional misadventures since the late 1980s, Pakistan is fighting a proxy war in the state of Jammu and Kashmir.[9] She has based her arguments based on the definition of strategic culture offered by Alastair Johnston, which defines strategic culture as an integrated system of symbols.[10] This definition highlights the role of the military forces in affairs of governance and articulating strategic preferences. The data source of Christine Fair is professional military publications and the research work is contemporary. I have based my arguments on the strategic culture of Pakistan as given by her. According to Dr Christina Fair, the Pakistan Army's strategic culture is characterised by four beliefs.[11] Firstly, that Pakistan is an insecure and incomplete state; second, that Afghanistan is a source of instability; third, that India rejects the two-nation theory and seeks to dominate and destroy Pakistan: and the last, that India is a regional hegemony that must be resisted. (sic) In this chapter, the Pakistani actions will be analysed in Afghanistan and against India and the hybrid war being launched by Pakistan against its neighbours will be explained with the help of a construct.

Pakistan's Interference in Afghanistan

Pakistani President Pervez Musharraf on his visit to Japan on 13 March 2002, urged Japanese corporates to come forward and participate in the economic revival of Pakistan which he said was important for stability in Afghanistan. Similarly, earlier on his visit to the USA, he had proposed that a stable Afghanistan along with Pakistan can provide the land-locked Central Asian countries the much-required gateway to the Indian Ocean.[12] This vision of President Musharraf was not new. In fact, this thought process is based on the geo-strategic importance of Pakistan's location and has influenced the strategic vision of the leaders of Pakistan since the early 70s. This vision has led Pakistan to detach itself from South Asia and align more towards the Islamic countries of West Asia.

Afghanistan from very beginning questioned the rationale of Pakistan, the legitimacy of the Durand Line and the division of Pashtun tribal areas.[13] This added to the insecurity of Pakistani leaders, who believed that India may seek to undo the Partition.[14] For newly-born Pakistan, the proposition of Afghanistan questioning the Durand Line based on division of ethnic identity, with support from India, became one of the biggest challenges.[15] In order to offset the Pashtun ethnic cauldron, Pakistan played the card of Islamic identity and ideology. It played this card well to blunt the call of ethnic nationalism being taken up by Afghanistan.[16]

The Pakistani rulers and intelligentsia believed that due to its geo-strategic location, it has the potential to play a leading role in the region.[17] Later, it was for the same reason that the USA chose Pakistan as the frontline to dislodge the Russians from Afghanistan. However, Pakistan had started interfering in the affairs of Afghanistan much earlier, may be in the 1960s, when Pakistan's intelligence agencies were encouraging Islamist political groups to pursue a forward policy of seeking ideological allies and setting up Islamic institutions in Afghanistan.[18] Influenced by Maududi and his Jamaat-e-Islami Pakistan, Burhanuddin Rabbani in Afghanistan founded the Jamaat-e-Islami Afghanistan in 1972.[19] It was Abul-Ala-Maududi and his Jamaat-e-Islami which provided shelter to Burhanuddin Rabbani in Pakistan when Muhammad Doud Khan, President of Afghanistan at that time ordered his arrest. Ahmad Shah Masood who led the military wing of Jamaat-e-Islami Afghanistan (largely comprised of Tajiks) and Gulbuddin Hekmatyar who later founded Hizb-e-Islami (primarily comprising of the Pashtuns), were followers of Maududi and later led the revolt against the Soviets and played a pivotal role in pushing them out of Afghanistan in 1979.[20] According to Hussain Haqqani in '*Pakistan: Between Mosque and Military*', an Afghanistan Cell was established in the Pakistan Foreign office as early as 1973. He has quoted a former principal advisor to Gen. Zia regarding the ISI conducting intelligence operations within Afghanistan.

Pakistan employed hybrid warfare in Afghanistan to fight the Soviets on behalf of the USA. Financed by the USA, Pakistan provided not only the manpower, training, weapons and ammunition, administrative and moral support, but also the most crucial aspect of military advice to the leadership of these Islamist radical groups.[21] Mujahideen trained and equipped by the ISI of Pakistan on behalf of one superpower were successful in defeating the other superpower. It was to the credit of the ISI which managed the separate group of tribes which have inconsistent loyalties and internal ethnic conflicts. It could mobilise such a diverse pool of mercenaries by carrying out religious indoctrination. This can be taken as a textbook example of the use of religion and radicalisation as part of hybrid warfare.

Pakistan did not stop interfering in Afghanistan's internal affairs even after the aim of the USA to send USSR back was achieved. Rather, after the fall of the Communist government of Nazibullah in 1993, Pakistan attempted to get a government which it could influence. Gulbuddin Hekmatyar, who broke off from the Jamaat-e-Islami Afghanistan, founded the Hizb-e-Islami (primarily comprised of Pashtuns), which managed to get the stronger backing of ISI and was operating from Pakistan. Rabbani suspected that Pakistan was supporting Gulbuddin Hekmatyar to seize power in Afghanistan.[22] Relations between Afghanistan and India under the rule of Rabbani improved. Due lack of strategic depth in the West, a pro-Indian regime in Afghanistan was the worst-case scenario for Pakistan. Hence, consequent to the failure of Gulbuddin Hekmatyar, Pakistan started looking for a new radical Pashtun group which would be more acceptable to the West, in particular to the Saudis, and can throw out the government of Burhanuddin Rabbani and Ahmed Shah Masood.[23] This led to the birth of the Taliban, which was conceived, delivered, nurtured, trained and radically indoctrinated by the ISI of Pakistan.

Taliban, meaning 'students of Islam', were largely drawn from Pashtuns, the largest ethnic group which represents 40 percent of the Afghan population. These young men of Islam were mostly young Koranic students recruited from madrassas which were set up in Afghan refugee camps in Pakistan. These madrassas radicalised and indoctrinated these young and vulnerable refugee children. Most of these madrassas were run by Maulana Fazlur Rehman and his Jamaat-e-Ulema Islam (JUI), which had considerable support from Pashtuns in Baluchistan and the North West Frontier Province (NWFP). Maulana Fazlur Rehman was a political ally of the ruling party in Pakistan who had direct access to the ISI, Army and Government.[24] Indoctrinated by the JUI and supported by the Pakistani establishment, the Taliban overran the country and in a matter of just two years controlled more than half of Afghanistan. Pakistan's close involvement was evident by the presence of their ambassador at Kabul and many ISI officers at Kandhar Shura, where Mullah Omar was crowned as Amir-ul-Momineed, or 'commander of the faithful'. While the arms, ammunition and vehicles were provided by the Pakistani establishment, the ideological guidance, training and manpower was being provided by JUI and the madrassas.[25] Pakistan was also the first country to provide political recognition to the Taliban government in Afghanistan in 1996.

The calculus changed for Pakistan in 2001, when, on September 11, the twin towers in the USA were attacked by Al Qaeda. Pakistan was left with no option but to participate in the global war on terror (GWOT) as the frontline state of the USA against its own progeny. The Taliban was pushed back by the Northern Alliance with the support of US forces. This was a huge setback for Pakistan as it was now losing the enormous leverage in Afghanistan, which it had with the Taliban regime.[26] Since, the fall of the Taliban and formation of a new government in Afghanistan, the Indian role and investment increased manifold. Apart from investing in social, infrastructure development and capacity-building projects like hospitals, libraries, roads and construction of the new parliament, India also increased its geostrategic reach by constructing Chahbahar port and Zaranj-Delaram road. The military cooperation between India and Afghanistan also increased with more than 1000 officers of the ANA being trained in India and also with assistance in military hardware. The government in Afghanistan signed a strategic partnership with India in 2011, while the same was rejected with Pakistan by the Karzai Government.[27] Pakistan continued playing a double game. On the one hand, it was the frontline state in GWOT and on the other, the Taliban had its safe havens in Pakistan. The recruitment and supply of weapons and ammunition kept flowing from Peshawar to Kandahar. The killing of Osama bin Laden in Operation Neptune Spear in May 2011 further exposed the dual role being played by Pakistan all along.

The President of the USA announced the first drawdown plan in December 11. Diversion of attention of the USA from Afghanistan provided the required window to the Taliban to further its military consolidation in Afghanistan under the shadow of Pakistan. By 2017-18, the Taliban and Pakistan were clear that the withdrawal of the US forces was eminent. This provided an opportunity to both the Taliban and Pakistan to regain control in Afghanistan. Now, in the renewed peace process in Afghanistan, all the stake holders like the USA, Russia, China and Iran are talking directly with the Taliban while bypassing the elected government in Afghanistan. This not only undermines the authority of the popularly-elected government of Afghanistan but also overemphasises the importance of Pakistan in the entire game, as the deep state in Pakistan still has a sizeable control on the Taliban.[28]

Pakistan's interference in Afghanistan over the last 50 years clearly indicates the application of various tenets of hybrid warfare. The use of religion to radicalise the youth, then training of irregulars, funding them and using them against the elected government fall in the realm of hybrid methods to achieve the national aim. All the major religious and terrorist organisations in Afghanistan like the Jamaat-e-Islami, Mujahideens, the Haqqani Network and the Taliban have been financed and supported either directly by or through the Pakistani establishment.

During the period of Soviet presence in Afghanistan, the funding for recruiting, arming, training and logistics of Mujahideen to fight the Soviets came from the USA. For creation of the Taliban, the initial funding was by Pakistan and later as the Taliban gained control of local administration, it developed its own resources.[29] The Haqqani Network's linkages with the Pakistani Army are well established. The Network has been receiving financial support from the Pakistani establishment. A declassified US Department of State cable suggests that the ISI paid $ 200,000 to the Haqqani network for carrying out the sensational terrorist attack that killed seven CIA operatives in Afghanistan in 2009.[30] Pakistan, through use of proxies like the Haqqani Network, Taliban and Jamaat-e-Islami, has been able to achieve the long term-aim of either to have a pro-Pakistan government in Kabul or keep Afghanistan in a state of instability.

Pakistan's Hybrid Warfare against India

Pakistan's strategy of engaging India with hybrid warfare has evolved over the years. It can be divided distinctly into two parts; first is the period from 1947 to 1980 and second from 1980 onwards. In the first part, Pakistan used the irregulars prior to launching conventional operations with the aim of exploiting terrain and local sympathy based on religion and fragile political situation. Post 1980, during the rule of dictator Gen. Zia-ul-Haq, there was a major shift in the strategy. Gen Zia, having learnt the lessons after losing three wars with India focused on exploiting the internal fault lines of India. The hybrid operations were not limited now only to the state of Jammu and Kashmir, but were targeted pan India, albeit the focus remained the Kashmir Valley.[31]

Period from 1947 to 1980. In this period, Pakistan has fought three major wars with India, i.e., in 1947-48, in 1965 and in 1971.[32] In all these wars, except in 1971, Pakistan has employed the tenets of hybrid warfare. The strategy followed was almost similar in the war of 1947-48 and 1965. The use of irregulars to shape the battlefield by exploiting the mountainous jungle terrain, harsh weather conditions local sympathy based on religious affiliations and poor infrastructural set-up. While conceptually the strategy may look very effective and promising and Pakistan did get success in the initial part of both the wars, but in the end Pakistan suffered defeat. A brief analysis of these two wars has been given below to bring out the lessons which contributed in transformation of the hybrid strategy of Pakistan against India.

1947-48 War. In 1947, Maharaja Hari Singh of Jammu and Kashmir had delayed his decision to join either India or Pakistan. Pakistan was getting anxious about the status of Kashmir. It felt that Kashmir is an essential and integral part in fulfilling its destiny as a homeland for Muslims in the Indian sub-continent.[33] In October 1947, thousands of Pakistani tribal Lashkars, equipped and organised by the Pakistani military, invaded the princely state of Jammu and Kashmir. The state forces were unable to put up any fight against the tribal raiders. Maharaja Hari Singh requested for assistance from India and agreed to sign the Instrument of Accession. The Indian Army airlifted troops to Srinagar airfield and was successful not only in stemming the progress of the raiders, but was able to push them beyond Poonch. The conflict ended on January 1, 1948, with the UN-sponsored ceasefire. The design of the Pakistan army to use tribesmen from the Western frontiers under the leadership of mostly ex-servicemen and a few serving military officers, equipped with weapons from its armoury is possibly the first recorded incidence of application of hybrid warfare by Pakistan.[34] The Pakistan military tried to characterise the war as an uprising by the Muslim majority against the Hindu ruler.[35] The civilian leadership, Prime Minister Liaquat Ali Khan and Mian Iftikhar who was minister for refugees and the finance minister Ghulam Muhammad were involved in the planning.[36] Brig. Akbar Khan, who was posted at Pakistani GHQ, wrote the appreciation (military paper written before a plan is made) for 'Armed revolt in Kashmir'. The plan involved use of tribesmen, former soldiers and a core of serving volunteer personnel and provisioning of 4,000 Punjab Police Rifles who were secretly diverted from the Pakistan Ordinance Factory.[37] Sardar Shaukat (Retd. Major) was appointed as overall in-charge, Khurshid Anwar (Retd. Major) as the commander of the Northern tribal force and Major Zaman Kiyani as commander of the Southern Axis.[38]

According to Agha Humayun Khan, the plan was envisaged at an opportune time but was poorly executed. The strategy and plan did not work due to a number of factors like poor ground leadership, over-reliance on the military acumen of the tribesmen, weak command and control structure and the most important being the timely and resolute action by the Indian Army. The Pakistani leadership also overestimated the religious affiliation with the Kashmiri Muslims.[39]

1965 War. The strategy of using tribesmen/irregulars to seize Kashmir was again used by Pakistan in 1965. The situation in 1965 in the context of both India and Pakistan was very sensitive. In Pakistan, military dictator Ayub Khan emerged victorious in the national election amidst allegations of rigging. This required him to distract the national attention from himself and convert public opinion in his favour.[40] The Pakistani leadership assumed that due to the debacle of the 1962 India-China war, the morale of the Indian Army was adversely affected and hence this was the best opportunity for Pakistan.[41] The tepid and half-hearted response of the Indian government in Kutch region against Pakistani intrusions gave the confidence to Ayub Khan to plan another audacious attempt to regain Kashmir.[42] The idea of initiating a guerrilla war against India in Kashmir has been in vogue in the minds of the Pakistani military since the 1950s. A secret "Kashmir Cell" was formed by the Foreign Office on Ayub's orders consisting of various key officials including the DMO.[43] The Pakistani military was emboldened and encouraged with the plan to ignite an insurgency in Kashmir, when the violent protests and riots took place after the holy hair of Prophet 'Moh-e-Moqadis' went missing for a short duration from Hazratbal shrine in Srinagar. The plan 'Operation Gibraltar' was put into action. It envisaged the induction of a large number of irregular forces into Kashmir to sabotage military targets, bridges, sever communication links, distribute arms to Kashmiri Muslims and eventually start an uprising against the Indian Government.[44] President Ayub was convinced that 'Operation Gibraltar' would succeed. In a written communication before the war, Ayub asserted that "*As a general rule Hindu morale would not stand more than a couple of blows delivered at the right time and place. Such opportunities should, therefore, be sought and exploited.*" (sic)[45] 'Operation Gibraltar' in Kashmir was to be followed up by 'Operation Grand Slam' in the plains sector of Akhnoor to cut off the state of Jammu and Kashmir from India.

The strategy of 'Operation Gibraltar' incorporated the lessons learnt from the failure of the 1947-48 war. The raiders in October 1947 plunged into loot, rape and large-scale destruction, which had alienated the local population. This time the existing unrest amongst the Kashmiris due to the disappearance of the holy relic was exploited. The infiltrating columns of raiders were much better organised in terms of command, control and communication. The Pakistan Army had received modern weapons and equipment from the United States of America as a result of a joint anti-Communist front, which qualitatively did improve its military capability.[46] The success of the entire operation based on the concept of hybrid warfare rested on three pillars—firstly, the successful exploitation of the existing unrest in Kashmir Valley; secondly, the success of infiltrating columns in cutting off the lines of communication by destroying bridges, destroying ammunition dumps and logistical installations and finally the success of a conventional attack in Akhnoor sector. According to J.N. Dixit, the plan was based on the operational tactics of the Peoples Liberation Army of China and Algerian guerrillas.[47] The operation was brilliant as a concept but the execution was not effective. J.N. Dixit has further highlighted that the infiltrators failed in their mission due to the resistance from the locals who passed on exact information about the location and movement of the infiltrators to the Indian Army.[48] The timely counter-action based on the information provided by the locals ensured that the first two pillars of the hybrid strategy of Pakistan collapsed in the initial stages itself.[49] Pakistan lost the 1965 war not only militarily but also in politico-strategy and policy objectives.[50] However, Pakistan's hybrid warfare strategy did surprise India in the initial part of the war.

Period Post 1980. As discussed earlier, the idea of initiating a guerrilla war against India in Kashmir has been in vogue in the minds of Pakistani military since the 1950s. Towards that end, a Kashmir cell was also created on the orders of President Ayub. However, this idea came to the forefront only during next military dictator's time, i.e., Gen. Zia, who having learnt the lessons from wars fought in 1947-48, 1965 and 1971, was convinced that Pakistan does not have the military capability to fight another conventional war with India.[51] He accepted the fact that Pakistan was no match to India either in military or in economic terms. However, at the same time the revenge of 1971 has always been an emotive issue for the Pakistan military. Gen. Zia brought in a paradigm shift in the strategy of Pakistan, particularly towards Kashmir and India as a whole. Lt. Gen. (Retd) Ata Hasnain calls it the Zia Doctrine. Gen. Zia focused on three aspects—first, to achieve a nuclear parity with India; second, to secure alliances with economically strong Islamic states using the leverage of nuclear status and third to wait for an opportune moment to exploit the internal fault lines of India like religion, caste, linguistics, ethnic and social.[52] Gen. Zia launched the war of a 'thousand cuts' against India. This was the fundamental shift in Pakistan's strategy towards hybrid war. While the focus was on Kashmir, simultaneously it involved a multi-pronged engagement of India through religious means, information warfare by unleashing vicious propaganda against the Indian government; economic warfare by flooding Kashmir with fake currency; cyber warfare by hacking and targeting the critical networks; psychological warfare by carrying out terror attacks at centres of economic, religious and security forces, along with conventional actions on the line of control but below the threshold of war; and while doing all this, keep playing the bluff of nuclear war. Since the 1980s, Pakistan has been fighting an informal war with India, albeit formally.

In the early eighties, one such fault line of India that was exploited by Pakistan was in the state of Punjab. Pakistan not only aided and abetted Sikh militants in Punjab but also exploited them for its own nefarious designs.[53] By early the nineties, the Indian Government successfully put an end to the Punjab unrest. Jammu and Kashmir, due to its geographical location, poor infrastructure set-up, religious affiliation, socio-economic condition and political instability proved to be an ideal battleground to test all the tenets of hybrid warfare. By 1987-88, Pakistan was able to initiate and fuel a separatist movement in Jammu and Kashmir. It began with the Jammu and Kashmir Liberation Front (JKLF) seeking independence for Kashmir with complete sponsorship of arms, ammunition, training, funding, moral and ideological support from Pakistan. However, soon, the JKLF lost steam as the Hizb-ul-Mujahideen (HM), a pro-Pakistani group, became the favourite proxy terrorist group of the ISI.[54] Dr. Fair in her scholarly work has covered in detail how Pakistan has created, cultivated, nurtured, supported, trained, financed and operationally employed them to achieve its national aim.[55] Pakistan created militant groups based on various ideologies within Islam like Deobandi groups of Jaish-e-Mohammad (JeM) and Harkat-ul-Ansar/ Harkat-ul-Mujahideen. The Jamaat-e-Islami influenced groups such as the Hizbul Mujahedeen and Al-Badr and Ahl-e-Hadith organisations like Lashkar-e-Taiba (LeT). The Pakistani media, in defence of the country, may call them Kashmiri organisations but on the ground, these militant groups just have very nominal representation of cadres from the state of Jammu and Kashmir and practically nil in leadership. All these militant groups have their headquarters and leaders in Pakistan. These militant groups not only operate pan India and Afghanistan[56] but have been involved in terrorist acts in Pakistan too.

In 1999, Pakistan initiated another hybrid war against India when the Pakistani military employed terrorists and troops from the Northern Light Infantry to occupy a number of high altitude posts in the Kargil-Dras sector of Jammu and Kashmir. Dr. Fair based on interviews with Pakistani officers involve in the planning of Kargil operations has argued that there were three principal reasons for the Pakistani military to undertake this misadventure. These were firstly, the Pakistani Army's desire to punish India for the 1971 defeat; secondly, to avenge the Indian Army's deployment at Siachen Glacier and lastly against India's shelling in the Neelum Valley sector and other such places.[57] The Kargil intrusions took place immediately after the very successful diplomatic initiative taken by the then Indian Prime Minister Atal Bihari Vajpayee, when he undertook a bus trip along the historical Amritsar-Lahore road link.[58] This raises the question that when Vajpayee took the initiative to resolve the India-Pakistan issue, then why did Pakistan undertake an action like intrusion in Kashmir? Professor Ganguli argues that the Pakistani military dominates all domestic and foreign policy decisions and a truce with India was the option that the military was looking for. In fact, this plan was possibly drawn by the Pakistan military as early as in 1996 and was aimed at taking revenge for the loss of Siachen Glacier. Whether the Pakistani prime minister was aware of any such plan, Professor Ganguli argues that even if he was briefed, it is doubtful how much of the implications he would have understood.[59] This further reinforces the psyche and strategic culture of the Pakistani Army towards any peace process with India. The Indian Army's resolute actions along with the diplomatic offensive got Pakistan isolated and eventually it had to vacate all the posts after suffering huge loss of men, material and reputation.

Post Kargil, the intensity of the violence by Pakistan-sponsored terrorist groups increased. The terror incidents were no more limited to the state of J&K. The nature of attacks like that on Parliament in 2001, the serial blasts in New Delhi in 2005, the 26/11 attack in Mumbai, the German Bakery blast in Pune, the attack on the Pathankot security base are a few examples which indicate the level of sophistication, availability of intelligence, coordination and audacity of the terror groups operating from Pakistan. This cannot be possible without the active support and sponsorship of the ISI. This chapter will now argue the various tenets of hybrid warfare that Pakistan has been using over the last 70 years through its proxies against India. Pakistan's hybrid strategy can be explained by use of the model given in Figure 3.1.

Classification of Means of Hybrid Warfare

The elements of hybrid warfare can be categorised in three parts. First being the conventional means like keeping the LC active by carrying out artillery shelling and small arms firing, occasional land grab actions or carrying out ambushes. The aim would be to exploit the no-war-no-peace situation on the LC and keep the Indian Army engaged on the LC and while doing so keep the activities well below the threshold of war. Second is the traditional sub-conventional means, i.e., use of terrorists to strike in depth on sensitive targets, like Parliament, Taj Hotel and German Bakery. The threat to use tactical nuclear weapons can also be considered a part of this. The third and most important is the new means like cyber, economic, religious, drugs, psychological and information warfare. Pakistan's use of these means simultaneously to keep India engaged can be qualified as hybrid warfare.

Figure 3.1: Means of Hybrid Warfare being used by Pakistan

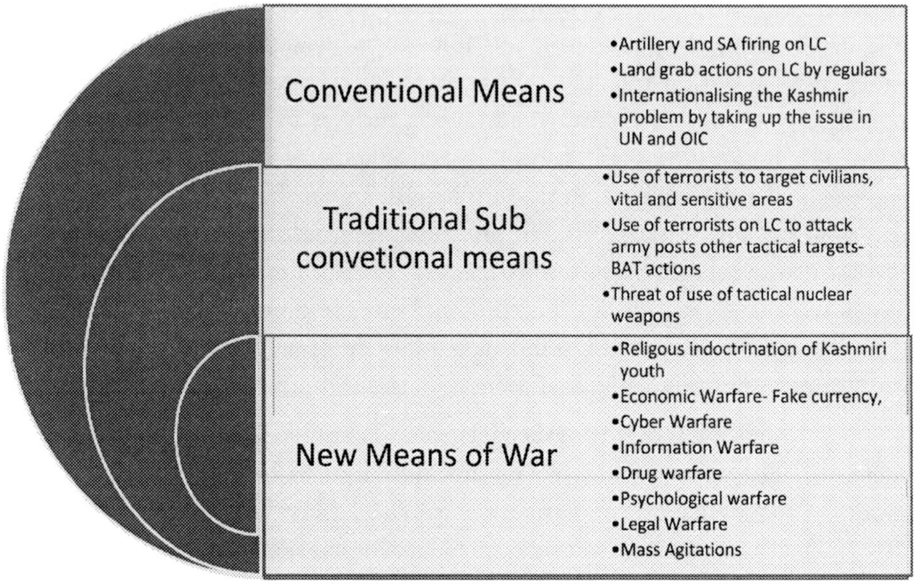

Source: Compiled from the chapter

Conventional Means. Pakistan uses conventional actions like firing across the LC, capturing of posts on the LC or occupation of an unhealed area on it as a part of improvement of defensive posture as a matter of routine to keep the LC active and engage the Indian Army. The details of the ceasefire violations in the state of Jammu and Kashmir is given in Table 3.1. Pakistan's diplomatic outreach to internationalize the Kashmir issue in the United Nations General Assembly and the OIC can also be categorized as conventional means.

Table 3.1: Cease-fire Violations in Jammu and Kashmir

Year	*Numbers*	*Casualties*	
		Security Forces	*Civilians*
2018	2936	26	28
2017	971	19	12
2016	499	13	13
2015	405	10	16
2014	583	08	14
2013	347	12	-
2012	114	05	-
2011	62	02	-
2010	70	04	-
2009	35	05	-

Source: Parliamentary questions and data from South Asian Terrorism portal[60]

Traditional Sub-Conventional Means. These means can be divided into two parts, the first being the use of terrorists/irregulars and the second the threat of use of tactical nuclear weapons. Pakistan has been targeting India through the use of terrorists as proxies and the issue of Kashmir is an alibi. This is well documented not only by Indian and Western scholars but even the Pakistani Military's official publication which endorses this philosophy. The theme of the Pakistani Military Green Book-2008 was 'Future Conflict Environment'. In the inaugural essay, Brig. Shaukat Iqbal has argued that the centre of gravity of Pakistan can be identified as 'Pakistan Military suitably backed by well-trained Mujahideen and reinforced by nuclear weapons'.[61]

- **Use of Terror/Terrorists/Irregulars.** Earlier in this chapter, Pakistan's use of terrorists groups as a state tool has been discussed. The terror attacks are generally carried out on high-value civilian targets like Parliament, five star hotels, at places which have large foreign footfalls and at crowded places which will create an emotional outburst like the Sarojini Market blast in Delhi in 2005.[62] Pakistan has also been using terrorists against military targets like Sunjwan[63] and Kaluchak[64] camps in Jammu, Brigade HQ in Uri,[65] Air Force base at Pathankot and on a number of times on convoys of security forces,[66] the last major one being the CRPF convoy at Pulwama.[67] The number of terror incidents recorded a sharp jump in 2018. The details of the terror incidents in the last five years are given in Table 3.2.

Table 3.2: Terror Incidents in Jammu and Kashmir

	2013	*2014*	*2015*	*2016*	*2017*	*2018*	*2019*
Terrorist Incidents	257	240	245	322	342	614	196*
Change in %	-	-	0.02%	0.31.4%	6.2%	79.5%	

*Inputs till 27 May 2019
Source: Parliamentary question for 2016, 2017 & 2018; SATP for 2013, 2014, 2015 & 2019[68]

Pakistan's strategy was based on not just convincing everyone that they were irrational enough to use nukes, but also being convinced themselves that India was rational enough to never risk calling Pakistan's bluff.

—Sushant Sareen

- **Threat of Use of Tactical Nuclear Weapons.** Pakistan's unabashed foray in conducting terror attacks in India can be connected to Pakistan having a nuclear weapon. India being a rational state always factored the possibility of any strong response escalating to nuclear war in the subcontinent and hence preferred the route of negotiations. Pakistan on the other hand being an irrational state, used nuclear weapons as an insurance policy and continued spreading terror in the subcontinent.[69] In 1999, during the Kargil war and later in 2002, consequent to the terror strike on the Indian Parliament on 13 December 2001 when the India Army was deployed and both nations were on the verge of war, India showed restraint in order to avoid a nuclear war. In India, defence experts converge on the possibility of the Pakistan military using tactical nuclear weapons in their own territory if the country is attacked. Sushant Sareen describes Pakistan's strategy as "***Pakistan's strategy was based on not just convincing everyone that they were irrational enough to use nukes, but also being convinced themselves that India was rational enough to never risk calling Pakistan's bluff.***" (sic)[70] However, this bluff of Pakistan was exposed when the Indian Army carried out a surgical strike on the terror camps located in PoK post the terror attack on Uri and, later when the Indian Air Force struck the terror camp at Balakot. In reaction to the determined Indian response against terrorists, Pakistan did not escalate the situation.

New Forms of War. The technological advancement added new dimensions to sub conventional warfare. These forms of warfare are more effective, more lethal, need lesser time to train and attain proficiency, practically no logistical requirement and very less economic investment. Hence, these have found favour amongst all the nations which are involved in sub-conventional warfare. As these forms of warfare are using new and emerging technologies, hence for ease of declassification they will be called as new forms of warfare. These are discussed below:

- **Ideological/Religious Warfare.** The Pakistan Army has used religion or the Islamic card as a unifying force and source of motivation to mobilise terrorist groups as a tool of foreign policy.[71] According to an *India Today* report, in the last 15 years or so the number of mosques controlled by Wahabis including the Ahle-Hadith has almost doubled. There are approximately 2,000 such mosques which are detrimental to the traditional Kashmiri Sufi shrines.[72] The funding for the Wahabi mosques comes through Saudi Arabia via the terror network/hawala channels from Pakistan. The increase in the number of madrassas and mosques is not restricted to the Kashmir Valley only. According to a *Hindustan Times* report, in the border areas of Rajasthan in the last few years the number of madrassas and mosques has increased substantially.[73] These madrassas and newly-constructed mosques have been successful in attracting youth towards the religious cause. This can be witnessed in the increased recruitment in militant groups from Kashmir. The madrassas and the schools by Falah-e-Aam Trust (FAT-educational organisation of Jammat-e-Islamic) have been instrumental in spreading radicalism. The increased radicalisation has led to increased recruitment by the terrorist groups. According to a report in *Hindustan Times*, the ISI's psychological operations have led to radicalization even among the well-educated and urban youth. In 2018, the local recruitment in terrorist groups has been over 200 whereas, in 2017 the figure was around 131.[74] The ISI has been persistently using all mediums including cyber to carry out propaganda of toxic, warped and intolerant religious and ideological narratives which are the root cause of weaning away the youth from Kashmiriyat towards radical Jihadism.[75]

- **Psychological Warfare**. The speed with which the spread of social media has taken place was beyond the imagination of any security analyst. Even before the security implications could be analysed, social media had become the most powerful tool of psychological warfare. Pakistan was quick to integrate the power of social media into its strategy of hybrid warfare. Special cells with cyber experts were created in Pakistan to create videos and other propaganda material to instigate violence and unrest in the Valley.[76] According to a report published in *The Tribune*, after Pakistan declared 2017 as 'Year of Kashmir', there has been a surge in videos showing "Sangbaz" (stone-thrower) with fiery songs calling for independence. The videos also carry "martyrdom appeals" to target the Indian security forces.[77] The report also argued that the video of a Kashmiri youth tied to an army vehicle to protect and evacuate the officers on election duty was shot professionally as the camera initially captures the youth from the front, then covers the view from the top and then from behind. It still remains a mystery as to who edited and uploaded the video. It was the work of professionally trained cyber warriors.[78] As per a report in *DNA*, in which it claimed to have the publicity material from Pakistan which gives out dates for conduct of cyber workshops for volunteers to fight a social media war in Kashmir against India.[79] By aggressive use of social media and propaganda, Pakistani agencies have been able to target the psychological domain of vulnerable Kashmiri youth. The videos of terrorists killed in action by security forces are circulated as martyrs with emotional songs in the background creating an overwhelming frenzy and emotional outburst amongst the youth. The crowd comprising young boys for stone pelting can be mobilised within a few minutes through an effective network of social media. The Directorate General of the Inter Services Public Relations (DG ISPR) in the recent Pulwama-Balakote episode was ahead in the OODA loop. DG ISPR was able to manage the narrative better and influence international perception. According to Lt. Gen. (Retd.) Ata Hasnain, DG ISPR has been successful in alienating Kashmiri youth from India.[80]

Figure 3.2

Source: Tribune, India[81]

- **Economic Warfare.** The International Narcotics Control Strategy Report (INCSR) is submitted annually to the United States State Department. In its 2011 report, it is highlighted that Pakistan is actively involved in pumping counterfeit currency into India. The fake Indian currency notes (FICN) are produced in Pakistan and pushed into India through the networks of terrorist groups and drug cartels that operate from Pakistan.[82] As per an RBI report, there was a 21 percent increase in the supply of FICN into India in 2014-15 as compared to 2013-14. As per the Financial Action Task Force (FATF), FICN was the ninth most circulated counterfeit currency in the world in terms of value and third most in terms of numbers. A report in *Diplomat* highlights the role of the ISI of Pakistan in distribution of FICN in India. The report quotes the testimony of David Hadley who was convicted for the Mumbai terror attacks in 26/11.[83] The same report also brings out the ISI connection with FICN based on the statement of Syed Abdul Karim Tunda who was arrested at the India-Nepal border in 2013. Tunda gave a detailed account of how he received a number of consignments of FICN from a serving brigadier of the ISI for distribution in India. The circulation of FICN has a dual purpose; firstly, to fund terrorist and terror activities, and secondly to affect the standing of the Indian economy.[84] The profit generated by the circulation of FICN is used in funding terror activities. Hence, it becomes a self-sustaining cycle (refer to Figure 3.3). The supply of FICN, apart from fuelling illegal activities has an adverse economic implication. It artificially increases the supply of cash in the market, thereby increasing inflation which over a period of time devalues the rupee, which adversely affects imports. Hence, FICN is low cost but a highly effective weapon of economic terrorism for Pakistan.

- **Diplomatic/Political Warfare.** Pakistan has been projecting itself as the crusader/champion of the Kashmir issue. Pakistani leaders have been consistently raising the issue of Kashmir in various international forums, the important ones being the General Assembly at the UN and the OIC. It has also been successful in forming a contact group on Kashmir in the OIC.[85]
- **Cyber Warfare.** India remains vulnerable to cyber attacks. According to the annual report available on the website of CERT, in 2018 alone more than two lakh security incidents related to cyber were reported and handled by CERT. The details of the last five years cyber security incidents are given in Table 3.3. As per the report of the CSIS, in March 2019, Pakistani hackers targeted more than 100 government sites.[86] Earlier in May 2018, the Pakistani military targeted government officials and other professionals of the Middle East, Afghanistan and India in an attempt to compromise them by inserting spyware in their computers which were distributed through Facebook Messenger.[87]

Figure 3.3: Self Sustaining Cycle of Terrorism

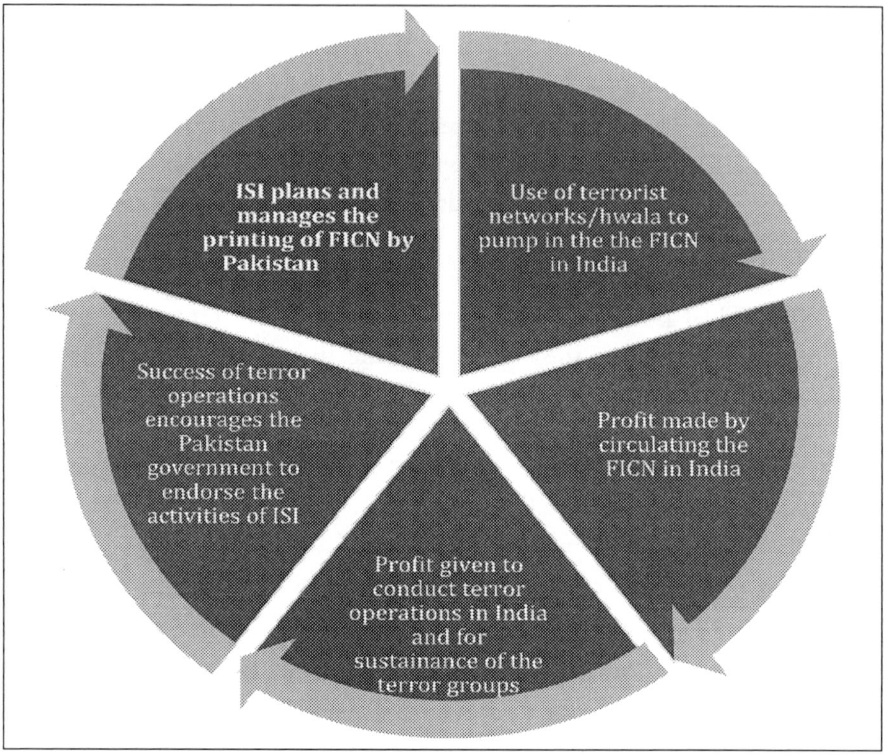

Table 3.3: Details of Cyber Attacks/Hacking of Websites

Sl.No.	*Year*	*Number of Cyber Attacks/ Hacking of Government sites*	*Reference of the PQ*	*Remarks*
1	2014	44679	Starred Question No 16, in Lok Sabha on	Cyber attacks
2	2015	49455	06.11.2016[88]	
3	2016	33147	Starred Question No 31,	Hacking of
4	2017	30067	in Lok Sabha on	Indian web sites
5	2018	15779 (till Nov 2018)	12.12.2018[89]	

Source: Figures taken from Parliament Questions

> "*Pakistan is attempting to flood Punjab with drugs as a 'long-term strategy' to destroy India's youth, particularly those living in Border States.*"[90] *(sic)*
>
> **—Capt Amarinder Singh**
> **Chief Minister of Punjab**

- **Narco Terrorism.** Pakistan has been targeting the youth of the border areas by pushing in drugs. The border areas in Punjab as well as in Jammu and Kashmir have been affected by narco-terrorism. This qualifies to be analysed as a secondary anti-India strategy to harm the youth of the border states.[91] Former DGP of J&K Shree S.P. Vaid considers drugs being smuggled into India from Pakistan as more dangerous than terrorism.[92] The infiltrating groups are tasked to carry drugs as part of their baggage with the twin purpose of funding terror acts and as part of the larger game of spreading the addiction. The year-wise details of recovery of drugs from the state of Jammu and Kashmir is given in Table 3.4.

Table 3.4: Provisional Data on NDPS recovered in J&K in 2018

Year	*Heroin (kg)*	*Ganja (kg)*	*Hashish (kg)*	*Opium (kg)*	*CBCS (bottle)*
2016	175	133.6	178.5	8.6	15.513
2017	207.5	210	33.3	0	17.799
2018	197	117.5	307.5	0.3	33.740

- **Legal Warfare.** Pakistan has been using various international forums to internationalise the Kashmir issue by bringing out false narratives. Repeated use of such false narratives by putting them under the legal apron is the hallmark of the Pakistani propaganda machine. Some of the case studies are given below:
  - The narrative of the Kashmir problem by the Pakistan Mission to the UN is full of factual inaccuracies, where it quotes the 'plebiscite' as given by the United Nations Security Council Resolution 47, adopted on 21 April 1948, concerning the resolution of the Kashmir conflict. It conveniently omits mentioning the first condition and the primary condition, that is, that Pakistan withdraw all its troops and nationals out of Kashmir.[93]
  - Pakistan acts as the champion for the rights of Kashmiris of the Indian state of Jammu and Kashmir, whereas the history of HR violations in PoK is actually the real concern. However, with repeated and consistent campaigns, Pakistan attempts to attract world attention once in a while, like in the case of use of non-lethal pellet guns by the security forces deployed in the state of Jammu and Kashmir. The collateral injuries due to the use of pellet guns by the security forces has been taken by Pakistan to the UN and eventually the Office of the United Nations High Commissioner for Human Rights

(OHCHR) in its report of June 2018 mentioned the use of pellet guns by the Indian security forces as an issue of concern.[94]

- There are allegations of Pakistan interfering in the administration of the state of Jammu and Kashmir through its proxies and OGWs who support terrorist groups. The ISI and other agencies provide the legal advice on how to exploit the liberal democratic system of India. One such case study is about the 'Roshni Act' in the state of Jammu and Kashmir. The original Act envisaged transfer of state land to its occupants, with some cost. Initially, 1990 was taken as the cut-off year for allocation of land.[95] However later the deadline was extended to 2004 and then to 2007. The Act has been misused by vested interests to change the demography of the state of Jammu and Kashmir.[96] Advocate Ankur Sharma, chairman of Ikk Jutt (Jammu) filed a PIL in the Jammu and Kashmir High Court with respect to allocation of land under the 'Roshni Act' citing it as an instrument of Jihadi terror and an effort to change the demography of Jammu.[97] The CAG report in 2014 also brought out the irregularities in the allocation of land under this Act. The State Administrative Council has repealed this Act w.e.f. 2019.[98] While further transfer of land has been stayed, the impact as questioned by advocate Ankur Sharma with respect to transfer of land to a specific community is an issue for further study. This may even be termed as an act of democratic terrorism, as this is a legal way to influence the electorate and eventually the outcome of elections.

- **Mass Agitation.** 2016 has been a watershed year in the Kashmir Valley. HM militant Burhan Wani changed the attitude in which the militants were looked upon earlier; he posted his photos on Facebook with an AK-47, which gave a new avatar to the local terrorists.[99] Post his killing, the entire Valley rose in revolt which carried on for more than seven months. His funeral was attended by more than 2 lakh people.[100] Pakistan grabbed the sentiment by declaring the day of the killing of Burhan Wani as a 'black day' for Kashmir and exploited the Kashmiri sentiment to win the elections in POK.[101] Since then, Pakistan has included mass mobilisation for organised activities like stone pelting and protests as tools of policy. The Pakistani establishment is using social media and proxies in terms of OGWs and local agents to ignite the emotions of the youth. NIA has traced a large number of illegal transactions in 22 accounts of South Kashmir leading to a huge amount of cash flowing into the Valley through Pakistan-based agents. The dates of these transactions could be co-related with the dates of major protests.[102] Earlier, the NIA had registered a case against JKART (Jammu Kashmir Affectees Relief Trust), a front outfit of the Hizbul Mujahedeen, which is sponsored by the ISI. As per the NIA, approximately Rs. 80 crore had been pumped into Kashmir by 2016 through JKART.[103] There have been reports of the use of cash for attracting minors for stone pelting, protestors to create obstruction in operations by security forces and throwing of grenades on security convoys.[104]

Table 3.5: The Pattern of Stone Pelting/Protests Hindering Functioning of Government and Security Forces

Year	*Number of Incidents (Stone pelting/protests)*	*Remarks*
2014	29	
2015	28	
2016	106	Burhan Wani killed
2017	205	
2018	142	
2019	53	Till May 2019

Source: The South Asia Terrorism Portal[105]

The two-nation theory, creation of Pakistan and persistent insecurity about its existence are some of the reasons which indicate the obsession of Pakistan with India. The unfulfilled desire of owning Kashmir; subsequent defeats in all the three wars fought to regain Kashmir; revenge for creation of Bangladesh and the desire to be seen and treated as equal to India by the world are factors that influence Pakistan's behaviour towards India. Haqqani brings out a thought-provoking comparison on the issue of parity. According to him "*The parity doctrine also requires Pakistanis to see India as an existential enemy. Textbooks still tell Pakistani children that Hindu India threatens Islamic Pakistan and seeks to terminate its existence. Hardly anyone outside of Pakistan believes that to be true.*" (sic)[106] As Pakistan has not been able to achieve parity in areas like democratic governance, education, economic and social development, subsequent governments have found it convenient to toe the line of the military, i.e., to try and achieve military parity. However, irrespective of the reasons, the methodology followed by Pakistan has a very high proportion of elements of hybrid warfare.

Analysis of Hybrid Warfare by Pakistan. Case studies of Pakistan's behaviour with Afghanistan and India bring out the enduring employment of hybrid warfare against its two neighbours. Traditionally, the use of irregulars or hybrid means is associated with a weak state against a superior power. However, the case studies of Afghanistan and India provide adequate evidence of how Pakistan applied hybrid warfare simultaneously against a weak power, i.e., Afghanistan as well as against a quantitatively as well as qualitatively superior power, i.e., India. In dealing with both the countries, while there were a few common threads like use of Islamic radicalism, creation of Mujahideens, spread of violence, conduct of psychological war by exploitation of the social media and information domain; at the same time, Pakistan has also calibrated the employment of hybrid warfare specific to each of the target nations based on respective internal fault lines.

Construct Showing Means of Hybrid Warfare and Intensity of Employment. The data given in this chapter on application of various means of hybrid warfare by Pakistan against India indicates how Pakistan has been able to manage the intensity of the means of warfare along with varying combinations of means. Thus, Pakistan has been able to control the conflict with India vertically in terms of intensity and horizontally in terms of combinations of different means of hybrid warfare. This is explained by the construct given below in two charts. These two charts capture a snapshot of the situation in the state of Jammu and Kashmir in 2014 and 2018. The snapshots of these two years have been taken as 2016 turned out to be a watershed in the evolving character of hybrid warfare in Jammu and Kashmir. Post the killing of Burhan Wani in 2016, the dynamics have changed in Kashmir substantially.[107] Figure 3.4, Chart 1 shows the situation as it was in 2104 and Chart 2 depicts the situation in 2018. In the charts, the X axis depicts the intensity and various forms of warfare are shown on the Y axis. In 2014, the intensity of cross-border firing, terror attacks, mass agitations were less than what we have seen in 2018. However, the economic warfare was a more effective tool in 2014 which after demonetization has come down in 2018. The religious, diplomatic and legal warfare has been kept at almost a similar level as these means are applied more at the administrative level. In the last few years, post 2016 Pakistan has exploited the local sentiments in Kashmir consequent to Burhan Wani's killing and heightened its hybrid warfare. The intensity has gone up in cross-border firing, terrorist attacks, psychological warfare, cyber warfare and narco terrorism.

Pakistan has employed the entire spectrum of hybrid warfare both against Afghanistan and India. In the conventional domain against Afghanistan, the

Figure 3.4: Construct Showing Means of Hybrid Warfare and Intensity of Employment

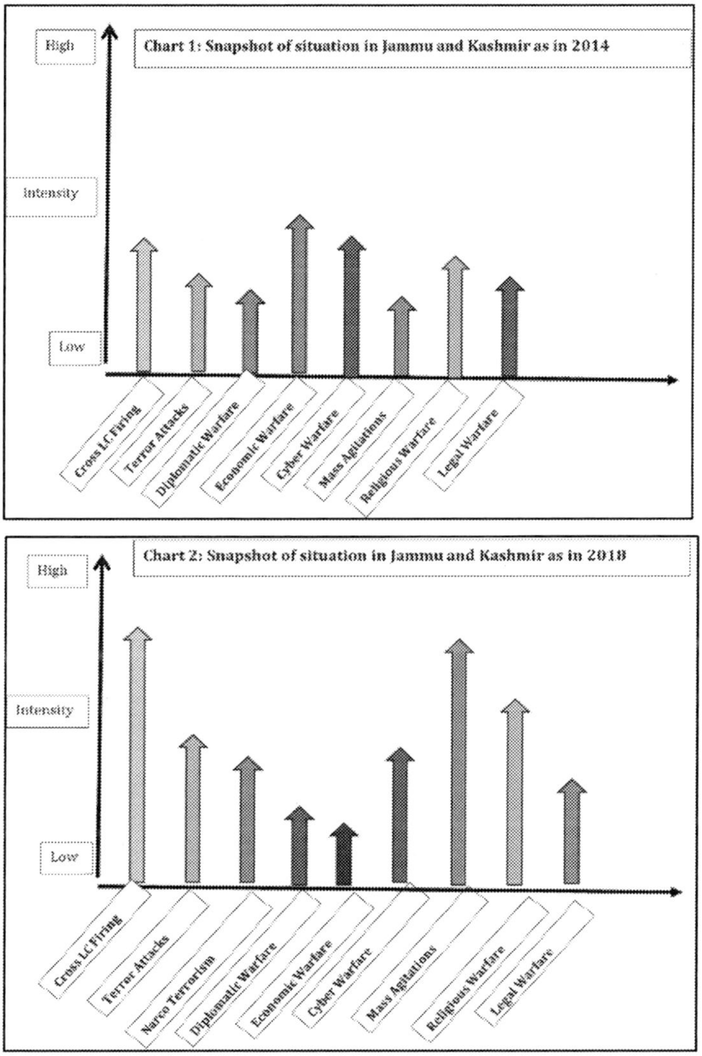

intensity has been limited to a few skirmishes on the Baluchistan border. However, against India, the employment of conventional means has been consistent. Apart from four wars, cross-border firing, BAT action on posts on the LC and infiltration have been routine phenomena, less a period of few years post ceasefire in 2004. The use of the Islamic card to organise and mobilise terrorist groups has been the common card with Afghanistan and India. The intensity has been up in the last few years for both the nations. On the diplomatic front, Pakistan has been able to seek international attention in the UN and especially at the OIC. On a few occasions, Pakistan has been able to get a few statements on human rights violations in Kashmir. Pakistan's success story lies in keeping itself relevant in Afghanistan, whether the Mujahideens were fighting the Russians in the eighties or against the US-led forces in the last few years. Whosoever has been fighting whomsoever in Afghanistan, the beneficiary has been Pakistan. The economic dimension of hybrid warfare has also been exploited by Pakistan with India and Afghanistan. The profit earned by printing and pumping of FCIN into India was used in terror funding, in organising mass protests, in stone pelting and for targeting convoys of security forces. In Afghanistan, opium cultivation has been the largest revenue earner for the Taliban. The money Pakistan received to manage the situation in Afghanistan was suitably used for reducing military disparity with India. Achieving nuclear capability was the watershed for Pakistan. Nuclear capability increased its status amongst the comity of Islamic nations; the monetary aid from Islamic nations which was earlier seen as charity now became an entitlement. The refusal to endorse no first use policy provided Pakistan the much required deterrence capability against any conventional operation by India. The new means of warfare like legal warfare, psychological warfare and cyber warfare are becoming more lethal and relevant with the advancement of technology especially in the information domain. The reach of the Internet and social media has altogether given a new dimension to the psychological warfare that has been used by Pakistan in the Kashmir Valley very effectively. The Friday prayers in mosques, the videos of funerals of terrorists, morphed pictures of atrocities on Muslims all over the world and live streaming of encounters are used for increasing mobilization of masses for stone pelting or preventing security forces to carry out their duties. The impact of psychological war in the hybrid domain has relegated the kinetic means to second priority to non-kinetic means. The development of information infrastructure in Jammu and Kashmir has unfortunately become a catch-22 situation for the Indian Government.

Conclusion

In both the case studies, the character of hybrid warfare has evolved as discussed in Chapter 2, technology being the most important driver. Pakistan has calibrated new technologies with traditional means of hybrid warfare. The training the terrorists in the use of the latest means of communication and the use of social media warriors is a case in point. Due to a higher technological threshold and development of communication networks in India, more components of hybrid warfare could be employed by Pakistan. The two case studies also indicate that in the last 70 years, Pakistan has been able to manage the situation in Afghanistan to its advantage. Even with India, while all its designs have failed, it has been able to keep India engaged by hybrid warfare. The driver behind the hybrid warfare by Pakistan is the strategic culture of the Pakistan Army. For the Pakistan Army, simply retaining the ability to challenge India is victory.[108] They have been very effective at ensuring that civilian institutions and even ordinary Pakistanis share the army's strategic concerns and priorities. The ensuing and enduring perpetual state of war with India and Afghanistan helped the Pakistan Army to keep itself relevant and in business. Whether the strategic culture of the Pakistan Army has benefited Pakistan in the last 71 years or not is a matter of subsequent study.

China has been successful in resolving most disputes in its favour or has maintained a position of advantage in ongoing disputes such as boundary issues in the South and East China Seas or against India. This has been achieved by China by exploiting the hybrid warfare and keeping the level of conflict below the threshold of war. In the next chapter, the Chinese actions in various disputes that it is managing have been analysed.

NOTES

1. Words said by the founder of Pakistan, Governor-General Muhamad Ali Jinnah speaking to guard commander in Peshawar, March 1948. Cited by Major General (Retd) Shaukat Ali Riza, *The Pakistan Army, 1947-1949*, Pakistan Army Book Club 1989, p. 305
2. Christine Fair in report to The National Bureau of Asian Research, Washington DC, on Pakistan's Strategic Culture. Available at http://www.nbr.org/publications/pakistan's-strategic-culture, accessed on 15 May 2019
3. Christine Fair in *Fighting to the End: The Pakistan Army's way of War*, Oxford University Press, p. 7
4. Lt. Gen. (Retd) Ata Hassnain, while speaking on "Kashmir : The Big Picture" at VIF, New Delhi, India
5. Bidananda Chengappa in "Understanding Pak's Strategic Culture Key." Available at https://www.deccanherald.com/opinion/perspective/understanding-pak-s-strategic-677531.html, accessed on 17 May 2019
6. Peter R. Lavoy in *Pakistan's Strategic Culture* prepared for Defence Threat Reduction Agency, USA
7. Ayesha Siddiqa in '*Military Inc: Inside Pakistan's Military Economy*' Oxford University Press, p. 13
8. Christine Fair in report to The National Bureau of Asian Research, Washington DC on Pakistan's Strategic Culture. Available at http://www.nbr.org/publications/pakistan's-strategic-culture accessed on 15 May 2019
9 Ibid
10. Ibid, see more at p. 5

11. Ibid, see more at p. 3
12. Ajay Behra in Pakistan's Strategic Vision: With and Without the Taliban, in Brookings, available on https://www.brokings.edu/opinion, accessed on 10 May 19
13. Haqqani, Husain (2005). *Pakistan: Between Mosque and Military,* published by Carnegie Endowment, p106
14. Ibid
15. Ibid
16. Ibid
17. Ibid
18. Ibid
19. Haqqani, Husain (2005). *Pakistan: Between Mosque and Military* published by Carnegie Endowment, p.171
20. Ibid, p. 173
21. Personal account by ex-ISI officer who was deeply involved in training the Mujahideen, written by Mohammad Yousaf and Mark Adkin, *Afghanistan: The Bear Trap, The Defeat of a Super Power*, New Delhi, Variety book Depot, 2006
22. Ahmad Bilal Khalil in The Tangled History of the Afghanistan-India-Pakistan Triangle available at www.academia.edu/the-tangled-history-of-the-Afghanistan-India-Pakistan triangle accessed on 13 May 19
23. Ajay Behra in Pakistan's Strategic Vision: With and Without the Taliban, in Brookings, available on https://www.brokings.edu/opinion, accessed on 10 May 19
24. Ahmed Rashid, *Taliban : Islam, Oil and the New Great Game in Central Asia,* London: I.B. Tauris, 2000, p.18
25. Ibid
26. Vivek Chaddha in Pakistan's Hybrid warfare in South Asia: Case study of Afghanistan and Pakistan, in *Hybrid Warfare and Changing Character of Conflict.* Published by IDSA, Pentagon Press, p. 50
27. Avinandan Choudhury in 'India in Afghanistan after Soviet Withdrawal', *Diplomat,* May 14, 2019, Available at https://.thediplomat.com/2019/05/ India-in-Afghanistan-after-Soviet-Withdrawal accessed on May 16, 2019
28. For a detailed analysis, refer to VIF issue brief 'Revival of Peace Initiative in Afghanistan: Implications for India' available at https://www.vifindia.org/brief/2019/january/31/revival-of-peace-initiatives-in-afghanistan-implications-for-india
29. Pakistani agents 'funding and training Afghan Taliban' in *BBC,* available at https://www.bbc.com/news/ 10302946 accessed on 15 May 2019
30. Lalit K. Jha, in ISI-funded Haqqani Network to attack CIA camp in Afghanistan available at https://www.indiatoday.in /pti-feed/story/ isi-funded-haqqani-network-to-attack-cia-camp-in-afghanistan-590412-2016-04-14 accessed on 17 May 19
31. Lt. Gen. (Retd.) Ata Hasnain, while speaking at IISS London on Discussion "Civil Society in Jammu and Kashmir: Democracy versus Terrorism" on 28 Mar 2019. Available at https://www.iiss.org/events/2019.03/ democracy-versus-terrorism accessed on 11 May 2019
32. The Kargil war of 1999 will be covered in Part II of Pakistan's hybrid war against India.
33. Christine Fair in *Fighting to the End The Pakistan Army's Way of War,* Oxford University Press p. 140

34. Vivek Chaddha in *Hybrid Warfare: The Changing Character of Conflict*, Pentagon Press, p. 39
35. Ibid
36. Agha Humayun Amin, The 1947-48 War in Kashmir: The War of Lost Opportunities, March 1999, available at https://archive.org/stream/The1947-48KashmirWarTheWarOfLostOpportunities/49202996-The-1947-48-Kashmir-War-Revised_djvu.txt, accessed on 20 May 2019
37. Parveen Swami in 'India, Pakistan and Secret Jihad', Asian Security Studies,, p. 20
38. Agha Humayun Amin, op. cit.
39. Vivek Chaddha in *Hybrid Warfare: The Changing Character of Conflict*, Pentagon Press, p. 40
40. Agha Humayun Amin, in The Grand Slam: The War of Lost Opportunities, March 2000 available at http://www.defencejournal.com/2000/sept/grand-slam.htm accessed on 22 May 2019
41. Vivek Chaddha in *Hybrid Warfare:The Changing Character of Conflict, Pentagon Press*, p. 41
42. Agha Humayun Amin in The Grand Slam: The War of Lost Opportunities, March 2000, available at http://www.defencejournal.com/2000/sept/grand-slam.htm accessed on 22 May 2019; Christine Fair in *Fighting to the End: The Pakistan Army's Way of War*, Oxford University Press, p. 92; J. N. Dixit in *India-Pakistan War and Peace*, Taylor and Francis, p. 146
43. Agha Humayun Amin in *The Grand Slam: The War of Lost Opportunities,* March 2000, available at http://www.defencejournal.com/2000/sept/grand-slam.htm accessed on 22 May 2019
44. Christine Fair in *Fighting to the End: The Pakistan Army's Way of War*, Oxford University Press, p. 93
45. Annexure-G to GHQ Letter Number 4050/5/MO-1 dated 29 August 1965. Directive from President Ayub Khan to General Mohammad Musa, Commander-in-Chief, Pakistan Army. Quoted by Agha Humayun Amin, in The Grand Slam: The War of Lost Opportunities. March 2000 available at http://www.defencejournal.com /2000/sept/grand-slam.htm, accessed on 22 May 2019; Bibliographical Notes—'*Zulfi Bhutto of Pakistan—His Life and Times*', Stanley Wolpert, Oxford University Press, Karachi-1993.
46. Read more about the 1965 War in *War Despatches; Indo-Pak Conflict 1965* by Lt. Gen.(Retd.) Harbaksh Singh, Lancer International, Delhi.
47. '*The plan was based on the operational tactics of the Peoples Liberation Army of China and the Algerian guerrillas. Once the initial operations succeeded, a revolutionary government would have set up a provincial government of free Jammu and Kashmir. This government was ultimately to be backed by two divisions of the Pakistani Army, masquerading as Kashmiri freedom fighters and guerrillas.*'(sic), see J. N. Dixit, in *India-Pakistan War and Peace*, Taylor and Francis, p. 138
48. J. N. Dixit, in *India-Pakistan War and Peace*, Taylor and Francis, p. 139
49. Vivek Chaddha in *Hybrid Warfare: The Changing Character of Conflict*, Pentagon Press, p. 42
50. J.N. Dixit, in *India-Pakistan War and Peace, Taylor and Francis*, p. 141
51. Lt. Gen.(Retd) Ata Hasnain, while speaking at IISS, London on Discussion "Civil Society in Jammu and Kashmir: Democracy versus Terrorism" on 28 March 2019. Available at https://www.iiss.org/events/2019.03/ democracy-versus-terrorism, accessed on 11 May 2019
52. Lt. Gen.(Retd) Ata Hasnain, while speaking at VIF, New Delhi on 'Kashmir the Big Picture' on 2 April 2019, available at https://www.vifindia.org/event/report/2019/april/02/talk-on-jammu-and-kashmir-the-big-picture, accessed on 21 May 2019

53. Christine Fair in *Fighting to the End: The Pakistan Army's Way of War*, Oxford University Press, p. 6
54. Vivek Chaddha in *Hybrid Warfare: The Changing Character of Conflict*, Pentagon Press, p. 43
55. For more details, see Christine Fair in *Fighting to the End: The Pakistan Army's Way of War*, chapter 9, Oxford University Press
56. Christine Fair in *Fighting to the End: The Pakistan Army's Way of War*, Oxford University Press, p. 243
57. Ibid, p. 151
58. Sumeet Ganguli in '*Deadly Impasse in Indo-Pak Relations at the Dawn of the New Century*', Cambridge University Press, 2016, p. 40
59. For more details, see Chapter 4 in '*Deadly Impasse in Indo-Pak Relations at the Dawn of the New Century*', by Sumeet Ganguly, Cambridge University Press
60. For details see https://www.satp.org/official-data/jammukashmir/cfa-violations, accessed on 02 Jun 2019
61. Christine Fair in Fighting to the End The Pakistan Army's way of War, Oxford University Press pp 42
62. 2005 Delhi Serial Blasts see https://www.hindustantimes.com/delhi-news/2005-delhi-serial-blasts-what-happened-that-day/story-SJsFozjWHKz8jHBQGILJvN.html, accessed on 22 May 2019
63. Sunjwan Army camp attack, see https://www.thehindubusinessline.com/news/national/operations-to-flush-out-terrorists-from-army-base-enters-day-2/article22720736.ece, accessed on 22 May 2019
64. 30 Killed in Kaluchak suicide attack, see https://www.thehindu.com/2002/05/15 /stories/2002051503030100.htm, accessed on 20 May 2019
65. Attack at Uri army camp, see https://www.hindustantimes.com/india-news/uri-terror-attack/ accessed on 21 May 2019
66. For more details about the terrorist attacks on security forces, see https://www.thehindu businessline.com/news/national/timeline-major-terror-attacks-on-security-forces-in-jammu-and-kashmir-since-1999/article26278836.ece, accessed on 22 May 2019
67. Terrorist attack on CRPF convoy in Pulwama, see https://www.indiatoday.in/india/story/pulwama-attack-2019-everything-about-jammu-and-kashmir-terror-attack-on-crpf-by-terrorist-adil-ahmed-dar-jaish-e-mohammad-1457530-2019-02-16, accessed on 22 May 2019
68. For details see https://www.satp.org/official-data/jammukashmir/cfa-violations accessed on 2 June 2019
69. Praveen Swamy in India-Pakistan and Secret Jihad, Routledge, 2007, chapter 7, pp 172
70. Sushant Sareen in Balakote air Strikes: The End of Madman Theory available at https://www.orfonline.org/research/balakot-air-strikes-the-end-of-the-madman-theory-48730/ accessed on 24 May 2019
71. Christine Fair in *Fighting to the End: The Pakistan Army's Way of War*, Oxford University Press, pp. 85-86
72. For more details, see, 'Jammu and Kashmir: Wahabi-controlled mosques on the rise, security forces tightening vigil', *India Today*, available at shmir-wahabi-controlled-mosques-ahle-hadith-sufi-shrines-militancy-madarsas-980513-2017-06-02, accessed on 24 May 2019
73. Increase in number of mosques and madrassas along the Pakistan border in Rajasthan, *Hindustan*

Times, available at https://www.hindustantimes.com/india-news/increase-in-number-of-mosques-madarsas-on-pakistan-border-in-rajasthan-bsf/story-bx8AJSntzO9BlOz QgKsHYK .html, accessed on 26 May 2019

74. How to counter radicalisation in Jammu and Kashmir, *Hindustan Times* available at https://www.hindustantimes.com/analysis/how-to-counter-radicalisation-in-jammu-kashmir/story-azWDdkR3o5zQa94wSWl2aL.html, accessed on 28 May 2019
75. Ibid
76. Pakistan Intensifies Cyber Warfare in Kashmir, *Tribune* available at https://www.tribuneindia.com/news /jammu-kashmir/ community/pakistan-intensifies-cyber-warfare-over-kashmir/395539.html, accessed on 24 May 2019
77. Ibid
78. Ibid
79. *DNA* Exclusive: Pakistan organizing cyber workshops to instigate unrest in Kashmir, available at https://www.dnaindia.com/india/report-cyber-war-is-the-new-face-of-insurgency-in-kashmir-2378034, accessed on 24 May 2019
80. Lt. Gen. (Retd.) Ata Hasnain, while speaking at IISS London on Discussion " Civil Society in Jammu and Kashmir: Democracy versus Terrorism" on 28 Mar 2019. Available at https://www.iiss.org/events/2019.03/ democracy-versus-terrorism, accessed on 11 May 2019
81. Pakistan Intensifies Cyber Warfare in Kashmir, *Tribune* available at https://www.tribuneindia.com/news /jammu-kashmir/ community/pakistan-intensifies-cyber-warfare-over-kashmir/395539.html, accessed on 24 May
82. Narayan Lakhman in Fake Currency from Pakistan threat to Indian Economy, *Hindu,* available at https://www.thehindu.com/news/international/Fake-currency-from-Pakistan-threat-to-Indian-economy-U.S./article14935520.ece, accessed on 24 May 2019
83. Sharmad Mahajan in India's Fight against Fake Currency, *Diplomat*, https://thediplomat.com/2016/11/ indias-fight-against-fake-currency/, accessed on 25 May 2019
84. Vivek Chaddha in *Hybrid Warfare: The Changing Character of Conflict,* Pentagon Press, p. 55
85. Ibid, p. 45
86. Significant Cyber Incidents available at https://www.csis.org/programs/cybersecurity-and-governance/technology-policy-program/other-projects-cybersecurity, accessed on 24 May 2019
87. Ibid
88. For more information on questions and answers, see Starred Question No. 16 in Lok Sabha on 06.11.2016, available at http://164.100.47.194/Loksabha/Questions/QResult15.aspx?qref=40475&lsno=16 accessed on June 29, 2019
89. For more information on questions and answers, see Starred Question No. 31 in Lok Sabha on 12.12.2018; Starred Question No. 16 on 06.11.2016, available at http://164.100.47.194/loksabha/Questions/QResult15.aspx?qref=73726&lsno=16 accessed on 29 June 2019
90. Capt. Amarinder Singh, Chief Minister of Punjab, "Pakistan wants to destroy youth of India' during Hindustan Times Leadership summit, available at https://www.hindustantimes.com/htls/pakistan-plans-to-destroy-india-s-youth-cm-amarinder-singh-on-punjab-drug-problem-at-htls-2018/story-bryVcksyokV1NvvE2KrSyJ.html, accessed on 26 May 2019
91. C.D. Sahay and Anurag Sharma in 'Jammu and Kashmir security review 2018—What lies Ahead,' VIF paper available at https://www.vifindia.org/brief/2019/april/15/jammu-and-kashmir-security-review-2018, accessed on 16 May 2019

92. S P Vaid, Former DGP J &K Drug menace bigger challenge than terrorism in J & K, available at https://economictimes.indiatimes.com/news/defence/drug-menace-a-bigger-challenge-than-terrorism-in-jammu-and-kashmir-dgp-sp-vaid/articleshow/61672910.cms accessed on 26 May 2019
93. For more see, 'Kashmir the History', as given on the web page of Pakistani Mission in UN available at http://www.pakun.org/kashmir/history.php accessed on 28 May 2019
94. OHCHR Report on the situation of human rights in Kashmir, available at https://www. ohchr. org/Documents/Countries/IN/DevelopmentsInKashmirJune2016ToApril2018.pdf accessed on 25 May 2019
95. For more details "The original Act" The Roshni Act envisaged the transfer of ownership rights of state land to its occupants, subject to the payment of a cost, as determined by the government. It was enacted by Chief Minister Farooq Abdullah's government, and it set 1990 as the cut-off for encroachment on state land. The government's target was to earn Rs 25,000 crore by transferring 20 lakh kanals of state land to existing occupants against payment at market rates. The government said the revenue generated would be spent on commissioning hydroelectric power projects, hence the name "Roshni"(sic) available at https://indianexpress.com/article/explained/jammu-kashmir-roshni-act-state-lands-satya-pal-malik-5475406/ accessed on 26 may 2019
96. Ankur Sharma, Chairman of IkkJutt(Jammu) while speaking in Indo-Analytics Conclave in 2019, available at https://www.google.com/search?q=ankur+sharma&rlz=1C1CHBF_enIN732IN732& source= lnms&tbm=vid&sa=X&ved=0ahUKEwjcnd-9q8jiAhXW8HMBHQMpCdkQ_AUIEygE&biw=1366&bih=625 accessed on 21 May 2019
97. The Roshini Act, Indian Express, available at https://indianexpress.com/article/explained/jammu-kashmir-roshni-act-state-lands-satya-pal-malik-5475406/, accessed on 26 May 2019
98. Ibid
99. For more details, see "2 yrs since Burhan Wani was killed: How Hizbul Commander changed the discourse of militancy in Kashmir," Outlook, available at https://www.outlookindia.com/website/story/how-burhan-wani-changed-the-definition-of-militancy-in-kashmir/313209, accessed on 26 May 2019
100. Ibid
101. Ibid
102. For more details, see "22 Bank Accounts in South Kashmir under lens of NIA for terror funding", *Times of India*, 26 August, 2016, online edition available at https://timesofindia.indiatimes.com/india/22-bank-accounts-in-S-Kashmir-under-NIA-lens-for-terror-funding-Hizb-links/articleshow/53865260.cms, accessed on 29 May 2019
103. Ibid
104. Ibid
105. Data on Stone Pelting and Protest in Jammu and Kashmir on South Asia Terrorism Portal, https://satp.org/other-data/india-jammukashmir/stone-pelting-incidents-jandk_2019, accessed on 28 May 2019
106. About 'Parity,' Hussain Haqqani writes, "Pakistanis are often not told of the widening gap between the two countries in most fields. For example, 94 per cent of India's children between five and 15 complete primary school compared with 54 per cent in Pakistan. Every year, 8,900 Indians get a PhD in the sciences compared with the 8,142 doctorates awarded by Pakistan's

universities since Independence. The total number of books published in any language on any subject in Pakistan in 2013, including religious titles and children's books, stood at 2,581 against 90,000 in India." For more, see,Hussain Haqqani in 'Pakistan in elusive quest for Parity', *The Hindu*, available at https://www.thehindu.com/opinion/lead/lead-article-pakistans-elusive-quest-for-parity/ article6845576.ece, accessed on 29 May 2019

107. For detailed analysis see, David Devdas in 'Generation of Rage in Kashmir' Oxford University Press

108. Christine Fair in *Fighting to the End: The Pakistan Army's Way of War*, Oxford University Press, p. 6

4

Chinese Characteristics in Hybrid Warfare and Its Manifestation

"To fight and conquer in all battles is not the supreme excellence; supreme excellence is in breaking enemy's resistance without fighting."

—**Sun Tzu in** ***Art of War***

ABSTRACT

In this chapter, Chinese white papers on defence have been analysed to study the evolution of Chinese military strategy and transformation of PLA and its convergence with hybrid warfare. The use of 'three warfare strategy and non-kinetic warfare' by China to achieve its national aims has been argued through case studies of 'Chinese coercion in South and East China Seas' and use of 'Economic Warfare'. Chinese hybrid warfare against India since independence and the likely threat in future has also been discussed in this chapter.

Chinese characteristics in hybrid warfare can be traced to Sun Tzu's classic '*Art of War*'. "*To fight and conquer in all battles is not the supreme excellence; supreme excellence is in breaking enemy's resistance without fighting.*"[1] This is reflection of oriental thought of violence being the last step towards achieving the national aim. Five thousand years later similar notes have been echoed in 1999 by two colonels in the People's Liberation Army of China, Qiao Liang and Wang Xiangsui, when they released a book on military strategy called *Unrestricted Warfare*, which as per Chinese translation reads as '*Warfare beyond Bounds*'. In chapter 5 of the book, they said, "*The great masters of warfare techniques during the 21st century will be those, who employ innovative methods to recombine various capabilities so as to attain tactical, campaign and strategic goals.*"[2]

In 2003, the Communist Party of China (CCP) approved and the Central Military Commission (CMC) adopted the concept of 'Three Warfares' which included 'public opinion, psychological and legal warfare'. A study for the Net Assessment Directorate, Office of Secretary of Defence of the USA, concluded that if the object of the war is to acquire resource, influence and territory and to project will—China's 'Three Warfare' is war by other means.[3] In the same study, Commodore Uday Bhaskar has argued "*that the core objective behind the Three Warfare concept is not to seek outright victory on any given matter, but to use its elements as a subtle Trojan horse means to realise the desired end.*"[4] The Chinese strategy is thus now going back to the philosophy of Sun Tzu, i.e., to win the wars without fighting.

Chinese Strategic Culture

The Chinese civilization being one of the oldest has profound historical, cultural and traditional foundations. Professor Huiyun Feng of the Utah State University in her book '*Chinese Strategic Culture and Foreign Policy Decision Making: Confucianism, Leadership, and War*' has observed that Confucian philosophy and thought has influenced Chinese strategic culture and has argued that it is primarily defensive.[5] Professor Huiyun Feng has argued, based on historical evidence, like the teachings of Sun Tzu, the construction of the Great Wall and restraint shown by the Chinese leadership during the Korean war-1950-53; Sino-Indian war of 1962 and Sino-Vietnam war of 1979 which amplify the defensive and peace-loving strategic culture.[6] This observation is extremely intriguing as it appears to counter the contemporary world view that China is a communist revisionist rising power which seeks to avenge the '*century of humiliation*'. The Chinese White Paper of 2005: 'China's Peaceful Development Road', published by the State Council Information Office of China, states, "*During the 100-odd years following the Opium War in 1840, China suffered humiliation and insult from big powers.*"[7] China's defeat by the British in the 1840-42 Opium War, followed by its defeat by Japan during the 1894-95 Sino-Japanese War, enforced unequal Boxer Protocol in 1901 and finally the second Sino-Japan war of 1937-45 were some of the key events which left eternal impressions on the Chinese people and had a profound influence in the shaping of China's modern strategic culture. More than 30 years of the unprecedented economic growth has been the driver for China's rapid rise in the hierarchy of global politics. The rising ambitions are a natural fall-out of the economic and military growth. In this chapter the Chinese White Papers on defence have been analysed to study the evolution of Chinese military strategy and transformation of PLA and its convergence with hybrid warfare. The use of the 'Three Warfare Strategy and non-kinetic warfare' by the Chinese to achieve their national aims has been argued through case studies of 'Chinese coercion in South and East China Seas' and use of 'Economic Warfare'. Chinese hybrid warfare against in India since independence and the likely threat in future has also been discussed in this chapter.

Analysis of the Chinese White Papers on Defence and Transformation of PLA

The first Chinese White Paper on Defence was released in 1998 by the Peoples Republic of China, which became a two-year ritual since then and the last one was released in 2015 called 'China's Military Strategy'. The analysis of these White Papers throws light on the thought process, according to which the PLA is being transformed. It also shows the increasing Chinese ambition and the role expected of the PLA. The causal relationship between assertive Chinese behaviour in managing boundary issues on the India border and in the South China Sea can be established through the analysis of the White Papers.

The initial few White Papers. i.e. till 2010, were more of generic and followed a similar format. However, they provided the foundation to Chinese strategic military thought. While covering the international security situation, defence policy, defence construction and arms control, the papers primarily stressed on 'peace' and 'development' as the 'major themes' of the relevant governments.[8] The concepts of 'Active Defence' and 'People's War' were introduced, which continue till date. The details of reduction in the size of the PLA by some 50,000 men were first time given out in 2000. The 'Revolution in Military Affairs (RMA)' through Chinese characteristics appeared in the paper of 2004 and the focus on modernisation of the Second Corps Artillery, PLA Navy and PLA Air Force to increase the PLA's war fighting capabilities and comprehensive national defence was also covered in the same year. The detailed defence modernisation plan till 2049 and also information about the PLA's 'organisation' and 'command and control' set up was given out in 2006, which also articulated the 'historic missions'.[9] The 2006 White Paper also summarised the reduction of PLA troops in 1985 (by one million), 1997 (by 50,000) and 2003 (by 200.000). It claimed the PLA strength to be 2.3 million in 2006. It also indicated the increase in the budget to the

Second Artillery, PLA Navy and Air Force by 3.8 percent and army reduced by 1.5 percent.[10] The 2010, defence White Paper did not have anything significant to indicate a change or major shift in policy.

In the 2013 White Paper the visible and significant change was the title itself, 'The Diversified Employment of China's Armed Forces', indicated the increasing confidence on the international stage due to its emergence as an economic superpower. It legislated that the modernisation of the PLA would be commensurate with China's international standing.[11] For the first time, China announced personnel numbers for the PLA Army's "mobile operational units" of 850,000, along with 235,000 for the Navy and 398,000 for the PLA Air Force.[12] These figures in isolation were deceptive as the strength of the Second Corps Artillery, People Armed Police Force, Militia, and Reserves was left out. The PLA personnel posted as the staff component in various formation HQs, the provincial and prefecture offices was also omitted. The White Paper also gave out the locations of all the 18 'Combined Corps' (Group Armies) under the Military Region, which again was new considering the Chinese characteristics. One important aspect that was omitted was the stance on employment of 'Nuclear Weapons'. Earlier White Papers covered this aspect under the 'Arms Control and Disarmament' section. Even the section on the defence budget was left out (was that by design) and needs further analysis as in 2013 there was a likely increase of 10.7 percent in the defence budget.[13]

The White Paper of 2015 with the title of 'Military Strategy', signalled a more coercive phase of China's rise. It indicated the resolve and determination to overcome any domestic or international impediment or obstacle to national development. This paper demonstrated a more focused explanation of China's strategic objectives and tasks for the military than the previous papers. It articulated that China's 'national strategic goal' was to 'complete the building of a moderately prosperous society' in all respects by 2021 and a 'modern, socialist country' by 2049. Well, these goals were not new.[14] However, for the first time they were reflected in the White Paper. The paper explained the requirement of the military to 'strive to provide a strong guarantee for completing the building of a moderately prosperous society in all respects and achieving the great rejuvenation of the Chinese nation'. This reinforced the idea that all military activities must support the pursuit of strategic objectives designed to realise China's potential as a prosperous, modern and powerful nation.[15] One of the important themes of the paper was the safeguarding of maritime and overseas interests. This upgraded the focus of the

PLA Navy from off-shore water defence to open seas protection.[16] Since then the relative allocation of budget has been higher for the Navy.

The paper also covered the new concept of 'holistic' security, which was introduced for the first time by President Xi Jinping at the Politburo study session in 2014. This concept integrates both domestic and international security. The paper enlarges the security spectrum to include 11 fields—political, territorial, military, economic, cultural, social, science, and technological, information, ecological, financial and nuclear. The enunciation of the holistic security policy in 2015 reflected the slowly settling realisation that the investment-manufacturing and export related growth for the last three decades had reached a plateau. Now to sustain continued economic growth, Beijing will need to execute extremely challenging policies in economics and trade, regional security order, governance and in the maritime domain.[17]

The White Paper also declared that China views control of sovereignty, territory and any other core interest non-negotiable (bottom line principle as announced by President Xi). It listed a range of security-related scenarios for which the PLA must plan. An analysis of the recent Chinese reactions to various situations with neighbours, at Doklam with India, at Senkaku Islands with Japan, Scarborough Reef with the Philippines and in the South China Sea show that the 'holistic security' concept facilitates China to take advantage of disputes. Military analysts in China believe that a military crisis handled properly provides a major opportunity to further national interests.[18]

The new security policy has several important implications as it legitimises the involvement of the military in a large number of issues beyond traditional security domains. Any issue of concern or crisis in any of the listed 11 areas which virtually encompasses the entire ecosystem of national functioning, will now can be covered under the security threat to the nation. The recent Chinese assertiveness in the South China Sea, on the India–Tibet border, trade, cyber and maritime domain in Xinjiang and most important towards the integration of Taiwan are reflective of the new security policy. According to some commentators, the Chinese military is likely to work harder and play a larger role in creating a favourable strategic posture.[19]

Transformation and Modernisation of PLA

Period till 2004. The modernisation of the PLA was one of the four pillars of the path breaking 'four modernisation program' under the charismatic leadership of Deng Xiaoping in 1978. In the sequence of priority, defence was fourth, after agriculture, industry, science and technology. The 1979 border war with Vietnam, in which the PLA had to struggle against the Vietnam army, was the first call to undertake reforms.[20] The technology threshold of Gulf War-I made the CCP leaders and the PLA hierarchy realise that in the present form, the PLA was unable to fight and win a modern war.[21] As the other three programs had made remarkable progress by then, the focus of the CCP now shifted towards modernisation of the PLA. The Military Balance 1996-97 gave the size of the PLA at 2,200,000 and when the all-active duty personnel were included, it was approximately 2,935,000. The PLA was organised into 24 Group Armies under the command of military regions.[22] Taking the lessons of the Gulf War 1, PLA embarked upon mechanisation of the PLA Army in 1997. At the same time, the first White Paper on defence was also released. After 1985, the first reduction of 50,000 troops was carried out in 1997.

The Second Gulf war of 2003 displayed the power of automation, networking, real-time information-sharing and the importance of situational awareness in speedy execution of operations. Military scholars and analysts carried out a detailed analysis of Gulf War II. The White Paper of 2004 covered the 'Revolution in Military Affairs (RMA)' through Chinese characteristics and the focus of reforms shifted towards modernisation of the Second Corps Artillery, PLA Navy and PLA Air force to increase the PLA's war fighting capabilities. The idea of comprehensive national defence was also covered in the same paper.

Reforms from 2005 till 2012 Onwards. In the White Paper on Defence of 2006 the Chinese government gave a three-step process for the further modernisation of the PLA. 'China pursues a three-step developmental strategy in modernising its national defence and armed forces, in accordance with the state's overall plan to realise modernisation. The first step was to lay a solid foundation by 2010; the second step was to make a major progress around 2020, and the third step was to reach the strategic goal of building informationalised armed forces and that were capable of winning informationalised wars by the mid-21st century.'[23] According to David Blasko, the PLA's operational goals in the middle of the first decade of the 21st century was to build capabilities for 'mobile operations and three-dimensional offence and defence' as it was graduating from 'regional defence to trans-regional mobility.' The focus was also to increase its capabilities in air-ground integrated operations, long-distance manoeuvres, rapid assaults and special operations. Blasko further qualifies that at that point of time for the PLA, the trans-regional mobility meant the capability to mobilise forces from one military region to another and along the coast; it was not related to the capability to project beyond the mainland of China.[24]

By 2012, some of the major reforms included consolidation of 11 Military Regions (MR) into seven Military Area Commands (MAC), conversion of army-intensive formations and the theatre command headquarters into a joint tri-service organisation and creation of distinct air force and naval headquarters.[25] The ground forces were further cut down to about 1.6 million, which was almost 27 percent (600,000) less than the 1997 figure quoted by Military Balance. The Group Armies were reduced to 18 from 24 in 1997.[26]

Table 4.1: Reforms in PLA: 1997 to 2012

Formation/Unit	*1997*	*2012*
Group Armies	24	18
Infantry Divisions	90	26
Armoured Divisions	12	04
Infantry Brigades	07	34
Armoured Brigades	13	14
Army Aviation Brigades/Regiments	0/7	4/9
SOF Groups/Regiments/Brigades	7/0/0	3/1/5

Source: David Blasko in PLA as Organisation V 2.0, in *PLA Ground Forces*, p. 263

Reforms in PLA Post 2102. The most far-reaching reforms and modernisation of armed forces have taken place under the leadership of Xi Jingping who became the Communist Party General Secretary in 2012. After taking over as chairman of the Central Military Commission, he created a Leading Small Group for monitoring of the transformation of PLA, of which he nominated himself as chairman. The first meeting of the group was held on January 27, 2015, and Xi went on to announce the reforms on 26 November 2015. In a short period of four months, the proposed reforms were discussed and approved by the CPC Politburo and CMC.[27]

Creation of Strategic Support Force: Capacity Building for Fighting Hybrid War. The Strategic Support Force (SSF), formed as a part of major reforms introduced in 2015, is an important milestone in the transformation of the PLA. The creation of the SSF demonstrates the importance and deep-rooted belief of Chinese strategic culture on information being the strategic resource in warfare. These reforms are aligned to the 'Holistic Security Policy' as given in the White

Paper on defence 2015, wherein the PLA sought to expand its role from land-based territorial defence to power projection in order to protect Chinese interests in the strategic frontiers of space, cyber space and the far seas.[28] the creation of the SSF centralises the space, cyber, electronic and psychological warfare capabilities which were under various services and departments of the PLA under one HQ.[29]

The SSF was created as part of a larger transformation from 2015 onwards, in which the erstwhile four all-powerful general departments were dissolved to form 15 joint service functional organs to function under the CMC. The SFF has been created by the brick model, wherein the operational units responsible for space, cyber and electronic warfare (EW) from the four departments have been reorganized. The overall structure of the SSF as per the missions is given below in Figures 4.1 and 4.2, respectively. The Strategic Support Force has two operational branches—the Space Systems Department (SSD) and the Networks System Department (NSD).

- **Space System Department (SSD).** The SSF's space mission falls under the SSD. The space capabilities of the PLA which were primarily under the General Armament Department and General Staff Department (GSD) have now been reorganised under this newly-created department. The SSD now comprises almost every aspect connected with space domain such as space launch and support, space telemetry, tracking and control, space information support, space attack and space defence.[30] The information support bases, the satellite main station (earlier part of GSD), and the satellite positioning main station which is responsible to manage China's Beidou satellite constellation has also been merged with the SSD.[31] This integration of all space-related support systems has increased the overall efficiency of the SSD.
- **Network Systems Department (NSD).** The SSF's cyber mission has been placed under the NSD. The NSD and its subordinate offices/forces are responsible for information warfare, including cyber warfare, EW and psychological warfare. The third department of the GSD, which was responsible for technical reconnaissance, cyber espionage and signal intelligence is the core of the NSD and the fourth department of the GSD which handled the cyber offensive force has been merged with the NSD.[32]

SSF and Three-Warfare Strategy. Prior to the reforms of 2015, the General Political Department (GPD) was responsible for political and psychological warfare. As part of the reforms, the SSF is now responsible for political and psychological warfare. The 311 Base is responsible for the conduct of political and psychological warfare against Taiwan.[33] It has six regiments under its control which form the core of China's psychological warfare forces in information operations.[34] The 311 Base is now part of the SSF Hence political and psychological warfare have also become part of the mission of the SSF. The, SSF now practically has all the elements/resources of information warfare to include cyber, communication, space, EW, political and psychological warfare. This reorganisation has achieved maximisation of resources, streamlined coordination issues and shortened the decision-making process.[35] As shown in Figure 4.2, the SSD and NSD have overlapping missions of counter space and strategic intelligence to facilitate coordination on common issues between the two verticals. The PLA, as a strategy relies heavily on psychological operations.[36] The integration of operational forces of 311 Base with technical reconnaissance, cyber espionage, signal intelligence and electronic emissions of the Third Department and the cyber warfare capability of the Fourth Department empowers the psychological operation forces with cross domain intelligence inputs. This reorganization will provide an integrated information support umbrella to the theatre commands during conventional operations as well as ongoing hybrid operations in the grey-zone period. The next part of the chapter covers the employment of hybrid warfare by China in the South and East China seas.

Figure 4.1: Overall Structure of the SSF

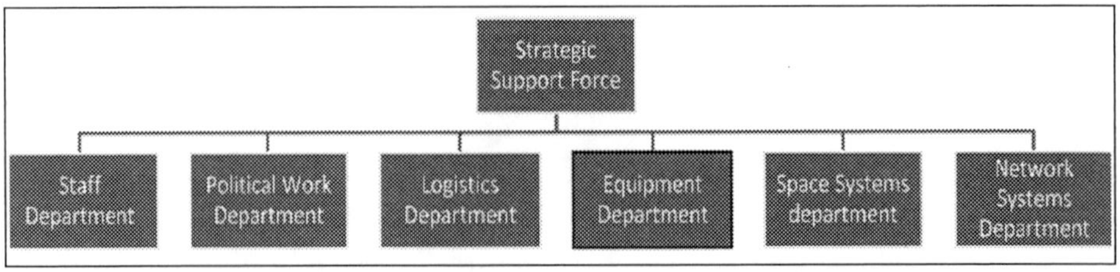

Source: Chinese Strategic Support Force, Centre for Study of Chinese Military Studies[37]

Figure 4.2: Overall Structure of the SSF according to Mission

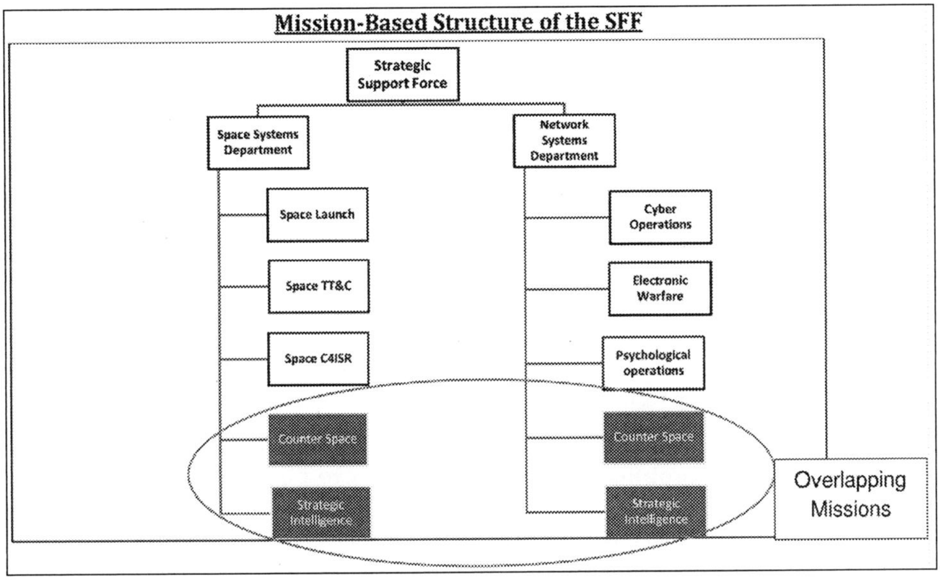

Source: Chinese Strategic Support Force, Centre for Study of Chinese Military Studies[38]

> *"China has continued to militarise the South China Sea (SCS) by placing anti-ship cruise missiles and long-range surface-to-air missiles on the disputed Spratly Islands and employing paramilitary forces in maritime disputes vis-à-vis other claimants. In the air, the PLA has increased patrols around and near Taiwan using bombers, fighters and surveillance aircraft to signal Taiwan. China additionally employs non-military tools coercively including economic tools during periods of political tensions with countries that China accuses of harming its national interests."*[39] *(sic)*
>
> **The Department of Defence**
> *Indo-Pacific Strategy Report*, June 2019

Securing Maritime Periphery through Hybrid Warfare

The above statement from the Department of Defence report on Indo-Pacific Strategy released on 1 June 2019 reflects the concern not only of the USA, but of all stake holders of the region. China's rise as an economic and military superpower has been the most exceptional story of growth and defining moment of the century in the world order. The increasing economic and military might has made China more confident, forceful and demanding. The Indo-Pacific region is increasingly confronted with assertive China, which is not hesitant in using economic, political and military coercion to secure national interests.[40] Though China dominates the Pacific Coast geographically, in order to achieve the great Chinese dream "*great rejuvenation of the Chinese nation by 2049, on completing 100 years of PRC,*" it needs to dominate the immediate Pacific Ocean. Moreover, in the last two centuries, all the invasions to China have been through the Pacific coast, i.e., by the Japanese and the British. The '100 Years of Humiliation' as quoted by President Xi was through the Pacific coast. In order to dominate the immediate Pacific, the area up to the second island chains becomes important. The Pacific coast till the first island chain consists of disputed territories, which potentially will restrict the access of the PLA Navy to the larger ocean. The First Island Chain being the line through the Kurile Islands (Japan), the Ryukyu Islands (Taiwan), the Philippines and Indonesia (Borneo to Natuna Besar) while the second island chain runs along a North-South line from the Kuriles through Japan, the Bonins, the Marianas, the Carolines and Indonesia. (Refer to Figure 4.3).

Figure 4.3: The Nine Dash Line or Alignment of First Island Chain

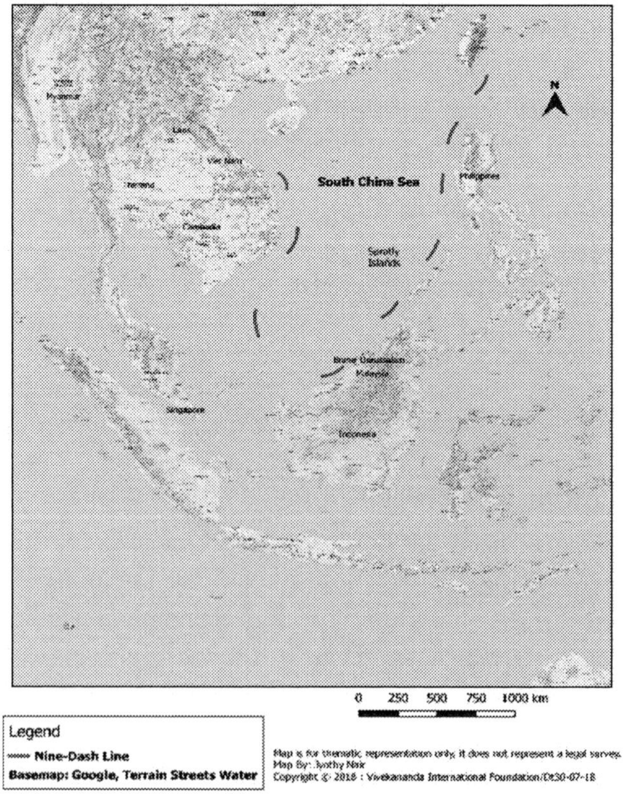

China has been pursuing its claim on the disputed territories of the first island chain through a variety of measures in an incremental and sequential way. It has so far refrained from using pure kinetic action or an all-out military operation. However, it has been using economic, political, diplomatic coercion to preserve its interests in the South and East China seas. This part of the chapter will argue that the Chinese actions in the South and East China region qualify to be hybrid warfare or the grey zone conflict as discussed in Chapters 1 and 2. As these disputes have a long and complex historical past, China has very astutely used a combination of legal warfare with economic and military means. In the past it has created situations or triggered a number of minor crises which have escalated the security matrix, then has gone for settlement of the crisis in its favour, thereby avoiding any escalation or conflict.

Conflict in the East China Sea

The East China Sea is surrounded by China main-land on the west, on the east are Japan's southern islands of Kyushu and Ryukyu and on the south are the islands of Taiwan. The East China Sea has been the nexus of Sino-Japanese distrust and strategic competition.[41] The Senkaku (as called in Japan)/Diaoyu (as called in China) islands comprise islets and three rocks in the East China Sea about 200 km north-east of Taiwan, 400 km west of Okinawa (Japan) and 300 km east of the Chinese main-land.[42] These islands cover a relatively small area of approximately. 6 sq km (Figure 4.4), However, it has been in the centre of a maritime dispute for the last 40 years because it is a potential reserve of oil, apart from its strategic importance.[43] China and Japan have based their claims on their

Figure: 4.4 Senkaku Islands

Source: South China Morning Post, 21 Feb 2019[45]

Figure 4.5: Activists on Senkaku Islands

Source: South China Morning Post

perception of historical records. In brief, while Japan's arguments are based on modern notions of international law and unchallenged administration, China's case is based on historical records and argues that these islands were a part of the imperial occupation and should have been returned at the end of World War II.[44]

The dispute started gaining prominence from 1990 when Chinese fishermen started entering the Exclusive Economic Zone (EEZ) of Japan. In 2004, seven Chinese activists landed on Diaoyu using fishing boats. These activists were arrested and later released due to diplomatic pressure by China (Figure 4.5).[46] In 2007, a Chinese research ship was sailing near the Senkaku islands; the Japanese protests were neglected by China on the grounds that the islands were part of Chinese territory.[47] On December 8, 2008, two Chinese government vessels intruded into Japan's territorial sea surrounding the Senkaku Islands. The ships remained in the disputed area for over nine hours even after adequate warning by Japanese navy patrol vehicles. The diplomatic protest also did not resolve the immediate crisis.[48] Such an act is an indication of intent to violate the sovereignty of another country and an attempt to change the status quo. In September 2010 a Chinese fishing trawler collided with a Japanese Coast Guard patrol vessel in the territorial waters of Senkaku Islands. After the collision, the Captain of the Chinese trawler was arrested by the Japanese Coast Guard.[49] Following the arrest, the Chinese government undertook a slew of economic and diplomatic measures. The economic coercion measures undertaken by China against Japan have been covered in detail in the later part of this chapter.

In 2012, the Japanese government nationalised the Senkaku Islands by purchasing them from their owners. Beijing reacted to this by increasing the intrusions into the territorial waters, imposing economic sanctions and consumer boycott of Japanese goods. The details of movement of the Chinese ships and other vessels in and around Senkaku islands are shown graphically in Figure 4.6. The graph has been sourced from the website of the Ministry of Foreign Affairs, Japan. It indicates the increase in the movement/intrusion by Chinese ships from 2012. Jeff Himmelman in his work, '*A Game of Shark and Minnow*' has quoted Maj. Gen. Zhang Zhaozhong of the PLA, who describes the PLA actions as cabbage strategy. The pattern of the PLA/Chinese movement around the Senkaku Islands may be compared to cabbage strategy.

> *'Cabbage strategy,' which entails surrounding a contested area with so many boats—fishermen, fishing administration ships, marine surveillance ships, navy warships—that "the island is thus wrapped layer by layer like a cabbage".*[50]

Figure 4.6: The number of Chinese government ships and other vessels that entered Japan's contagious zone or intruded into territorial seas surrounding the Senkaku Islands

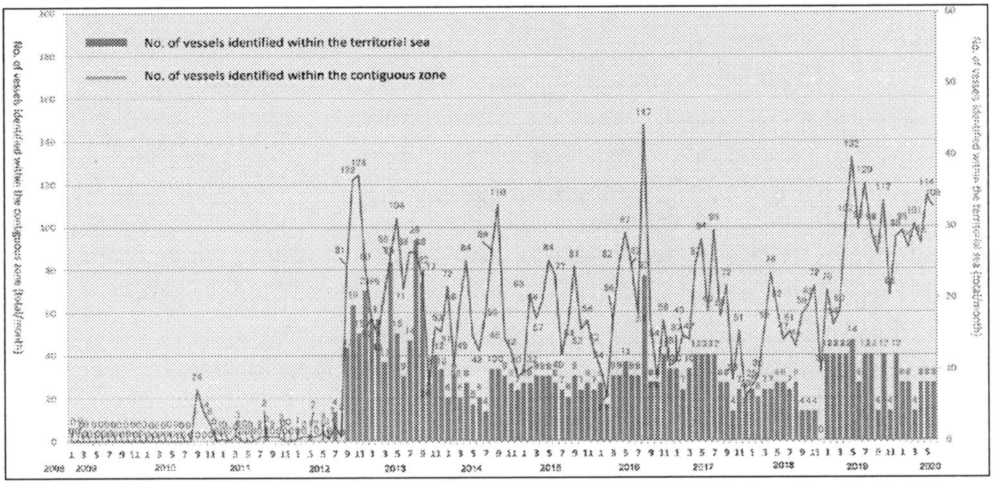

Source: Trends in Chinese Government and Other Vessels in the Waters Surrounding the Senkaku Islands, and Japan's Response[51]

In 2013, China escalated the situation by bringing in the dimension of movement in airspace. China declared Air Defence Identification Zone (ADIZ) over a part of East China Sea including the Senkaku/Diaoyu islands. The proposed ADIZ overlapped with the ADIZ of Japan, Taiwan and South Korea.[52] In recent years, China has also increased its air activities in and around Japanese airspace. The table below indicates the data on scrambles between Chinese and Japanese aircraft in Japanese airspace till 2015. In 2016 the number of scrambles against Chinese aircraft was 861.[53] (Figure 4.7)

Figure 4.7: Number of Scrambles against Chinese aircraft

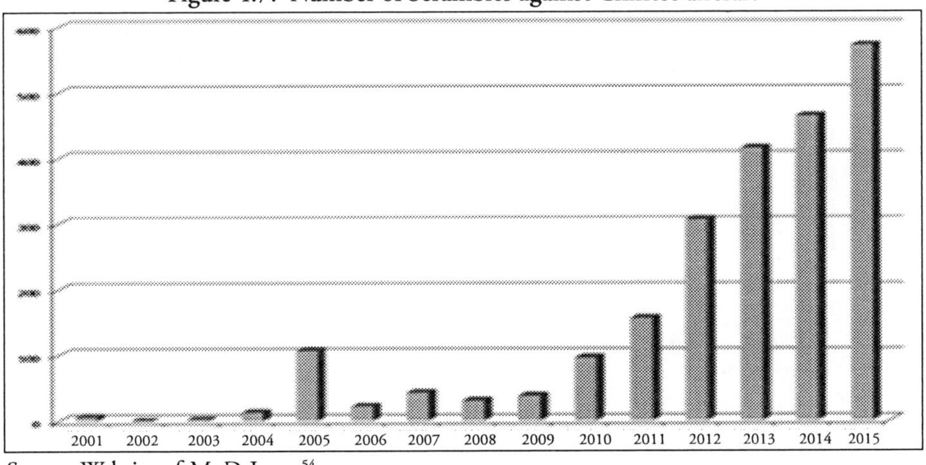

Source: Website of MoD Japan[54]

China has kept the level of hostile activities below the threshold of conflict in the East China Sea, while keeping the issue alive. It has used measures of economic and diplomatic coercion, created a false narrative by twisting historical facts and created a psychological pressure by using military measures like intrusions in naval and airspace. China, by using the hybrid technique, has been able to maintain an image of a responsible stake holder in an apparently genuine dispute and at the same time has had success in internationalising the issue.[55]

Chinese Activities in South China Sea (SCS). The South China Sea is the other region of economic and strategic importance where China has successfully employed hybrid warfare strategy to gradually but consistently and in a sequential manner to increase its control over disputed islands/reefs. Almost six countries stake overlapping claims in the area of the SCS. The area being contested is rich in resources like hydrocarbons and natural gas and is one of the major sea lines of communication. According to a report in 2016, the total value of trade through the SCS was 3.4 trillion USD. Five countries depended on the SCS for more than 50 percent of their trade.[56] Most of these countries are stakeholders/claimants in the territorial dispute in the SCS. According to a report by the Council on Foreign Relations (CFR), the region had approx. 11 billion barrels of oil, 190 trillion cubic feet of natural gas and approximately. 30 percent of global oil trade passage through this region in 2016. (Figure 4.8)[57]

Figure 4.8: Volume of Trade in South China Sea (in %)

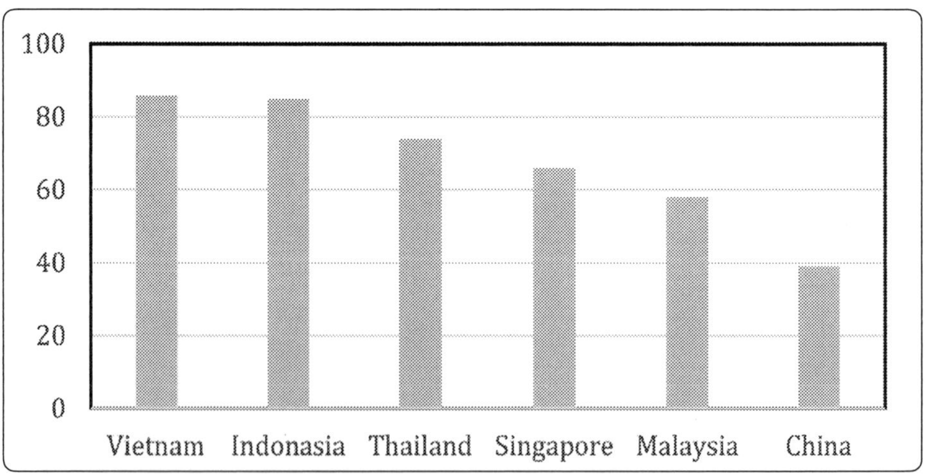

Source: Compiled from report in *The Diplomat*[58]

Maritime territorial disputes in the SCS span centuries.[59] The overlapping Exclusive Economic Zone (EEZ), natural resources, globalisation and increasing stakes of outside powers like the USA have periodically led to diplomatic tensions as well as confrontations. The dispute in the South China Sea primarily involves a group of three islands/shoals. These are explained in Figure 4.9.

- Paracel Islands are which are currently occupied by China are claimed by China and Vietnam.
- Spratly Islands are claimed entirely by China, Taiwan and Vietnam, and in part by The Philippines, Malaysia and Brunei, and are currently occupied in part by all these countries except Brunei.
- Scarborough Shoal is claimed by China, Taiwan and the Philippines and since 2012 is being controlled by China.

Figure 4.9: Maritime Dispute in South China Sea

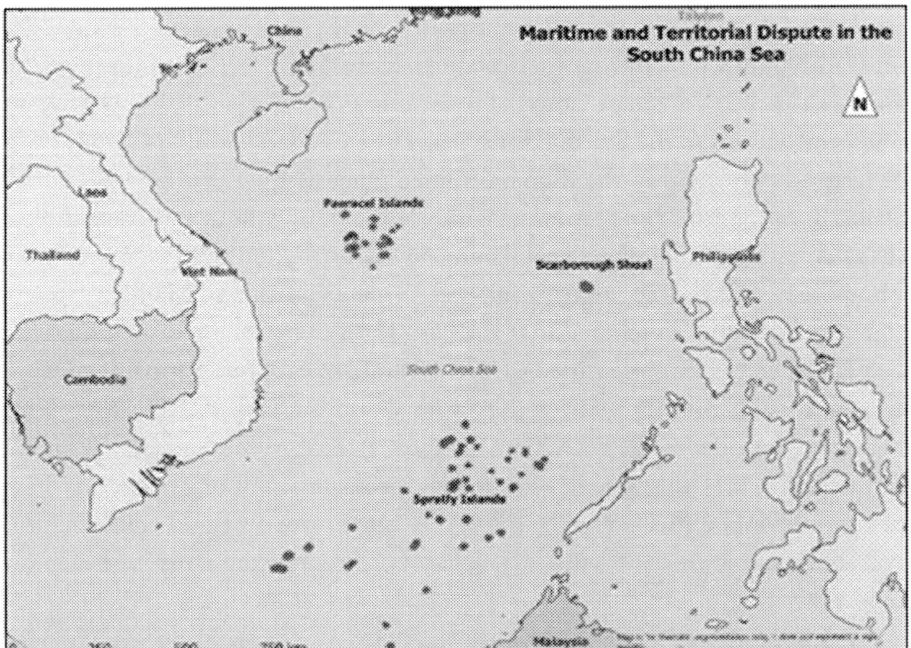

The Chinese strategy in the SCS is to protect its interests and to strengthen its claims and position it to employ the complete range of instruments of national power that includes diplomatic, informational, economic, military, paramilitary/ law enforcement, legal and civilian elements. Ronald O'Rourke in his Congressional Research Service (CRS) report on '*China's Actions in South and*

East China Seas: Implications for U.S. Interests—Background and Issues for Congress' classifies the Chinese strategy in the South China Sea as '***Salami Slicing Strategy***', that employs a series of incremental actions or of creeping annexation or creeping invasion, or as a "talk and take" strategy, in which meanwhile, on one hand, China engages in negotiations and simultaneously initiates actions to occupy the contested areas.[60] China has been increasingly using military, civil and non-government assets to deny other nations the legitimate use of resources in the in the South China Sea, including their own EEZ.[61] China twisted historical facts to generate its own narrative and documents in support of its claim on the disputed territories. Beijing's so called "nine-dash line," is a claim that encircles as much as 90 percent of the South China Sea. Beijing claims any land or features within this line based on a very vaguely defined 'historical maritime rights'.[62] This claim has been contested by Brunei, The Philippines and Malaysia based on proximity and UNCLOS (United Nations Convention on Laws of Sea) as it overlaps with their respective EEZs. As China is a signatory to UNCLOS, it has kept the 'nine-dash line' ambiguous and has never explained it explicitly.[63] The Philippines had approached the Permanent Court of Arbitration (PCA) against the assertive and unilateral actions in the South China Sea. The PCA declared that China's claims of historic rights within the nine-dash line, which Beijing uses to demarcate its claims in the South China Sea, were without legal foundation. The court directed that Beijing's activities like illegal fishing and construction of islands within The Philippines' exclusive economic zone (EEZ), are a violation of Manila's sovereign rights.[64] The court concluded that the land features in Spratly do not qualify to be islands, hence are not entitled to be maritime zones. The decision of court was seen as a major turning point is the South China Sea dispute. However, China rejected the verdict by the PCA, the Chinese president Xi Jinping post-declaration of the verdict said that China's "territorial sovereignty and marine rights" in the Seas would not be affected by the ruling.[65] China's utter disregard of international conventions, of which it itself is a signatory, indicates the Chinese hegemonic tendency.

China's creeping invasions of the disputed islands/shoals/reefs have been based on a few imperatives which are typical characteristics of hybrid warfare, such as the use of stealth to occupy the unoccupied islands or, if required, by minimum force develop them as their logistics hubs and as a base for further power projection. The option of armed confrontation is the last one and only to be taken when the situation is favourable.[66] The belligerent and aggressive actions taken by China are given below.

- **Island Building and Base Construction (2013 onwards).** China's land reclamation or the development of the island in the Paracel and Spratly Islands in the SCS has raised concerns among the countries in the region. China is developing military bases on six of these newly reclaimed islands. Major construction activities like airfields, ports and buildings have been noticed at Fiery Cross Reef, Subi Reef and Mischief Reef.[67] The development of such military capability provides China domination over the South China Sea. It also creates a psychological pressure on the smaller countries in the region. The pictures of the Fiery Cross islands before and after the construction have been carried out[68] are given below.

Figure 4.10: Fiery Cross Islands before and after construction.

Before Aug 2015

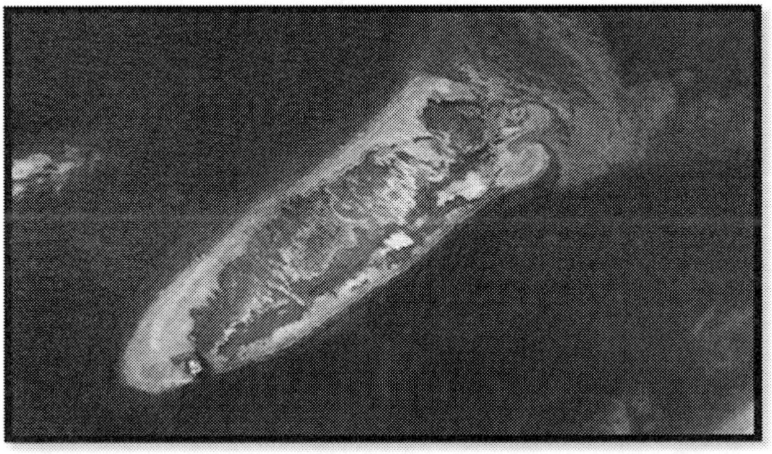

Sep 2015

Source: New York Times[69]

- **Declaration of ADIZ over South and East China Seas.** On November 23, 2013 China declared ADIZ over a portion of the South China and East China Seas. This is a belligerent act and against international norms.[70]
- **Implementation of Local Laws in International Waters.** On January 1, 2014, the fishing regulations administered by China's Hainan province were made applicable to waters constituting more than half of the SCS. Under this law any non-Chinese fishing boat could be apprehended and tried.[71]
- **Demographic Invasion.** China started inhabiting some of the disputed islands with Chinese nationals.[72]
- **Employment of Maritime Militia and Coast Guard Ships.** China uses its Coast Guard ships and maritime militia in maritime sovereignty-assertion operations. It has the largest Coast Guard fleet and uses it for the defence of its land.[73]
- **Scarborough Shoal Standoff.** In April 2012, the Philippine Navy discovered Chinese fishing trawlers and fishermen involved in illegal fishing close to the Scarborough Shoals. While the Philippine naval ship attempted to arrest two Chinese surveillance ships identified as Zhonggou Haijian 75 and Zhonggou Haijian 84, the latter positioned themselves between the Philippine warship and the Chinese fishing vessel, preventing the arrest.[74] Later, China used a combination of diplomatic, economic and military coercion to physically occupy the Shoal (the economic coercion has been covered in detail under Chinese economic warfare). The Philippines approached the Permanent Court of Arbitration (PCA) which ruled in its favour. However, China refused to accept or act on the PCA ruling.
- **China–Vietnam Ships Collide in South China Sea.** China established an oil rig in close proximity to the Paracel Islands claimed by Vietnam. The incident is also known as the Hai Yang Shi You 981. Vietnam despatched naval vessels in an attempt to stop China from establishing an oil rig in contested waters near the Islands. The encounter quickly escalated as China despatched 40 ships to protect the rig, resulting in a collision between several vessels in which Vietnam lost a ship.[75]
- **Strong Arm Tactics in South China Sea.** Chinese marine militia has been harassing fishermen from Vietnam and the Philippines by claiming the South China Sea as its territorial waters. Details of a few such reports

are given in Table 4.3. These are pressure tactics used by China to deter fishermen of smaller countries to venture into the South China Sea.

Table 4.2: Chinese Action against Fishermen in South China Sea

Month/Year	*Affected Country*	*Incident*
Jan 2005	Vietnam	Chinese Coast Guard boats fired on Vietnamese boats, killing 9 fishermen[76]
2005 to Mar 10	Vietnam	A large number of Vietnamese boats were attacked by Chinese maritime militia/Coast Guard boats[77]
Mar 2012	Vietnam	China arrested 21 Vietnamese fishermen[78]
2014	Vietnam	China arrested 6 Vietnamese fishermen[79]
Apr 2017	Philippines	Chinese Coast Guard fired on Philippine fishermen[80]
Jun 2017	Vietnam	Chinese boat attacked Vietnamese fishermen[81]
Mar 2019	Vietnam	Chinese marine surveillance boat chased Vietnamese fishermen's boat, which collided with rock and sunk[82]

Source: Compiled from the chapter, references are given in third column of the table.

Analysis of Chinese Actions in East and South China Sea. The Chinese interests, claims and assertive behaviour in the South and East China seas have been growing incrementally and gradually. In the East China Sea, initial simmering was observed in the 1970s and slowly the frequency of incidents increased in the first decade of the 21st century and the intensity and sensitivity increased after 2012. Similarly, the incidents in the South China Sea started with claims, then graduated to prophylactic firing to shooting on poor fishermen. The intensity increased with naval ships confronting each other and today it has reached to the declaration of an Air Defence Identification Zone (ADIZ) and establishment of military bases on artificially created islands. This gradual coercion or incremental coercion can be established imperially. The picture indicates how the gradual coercion by China has altered the dynamics of the South China Sea. China through a methodological way has been increasing the tempo of activities and has gradually converted the claims into a permanent position.

Chinese hybrid warfare in the South and East China seas can be essentially broken down to four elements for the purpose of analysis. The first and foremost is the use of 'Three warfare strategy', i.e., legal warfare by exploitation of existing rules to introduce own rules and regulations; twisting historical facts and maps to develop a new narrative to create a crisis; occupation of islands in disputed areas and then exploit the international law of EEZ to harass and prevent others to carry out any economic activity; applying Chinese laws like application of Hainan

Figure 4.11: Chinese Gradual Coercion in South China Sea

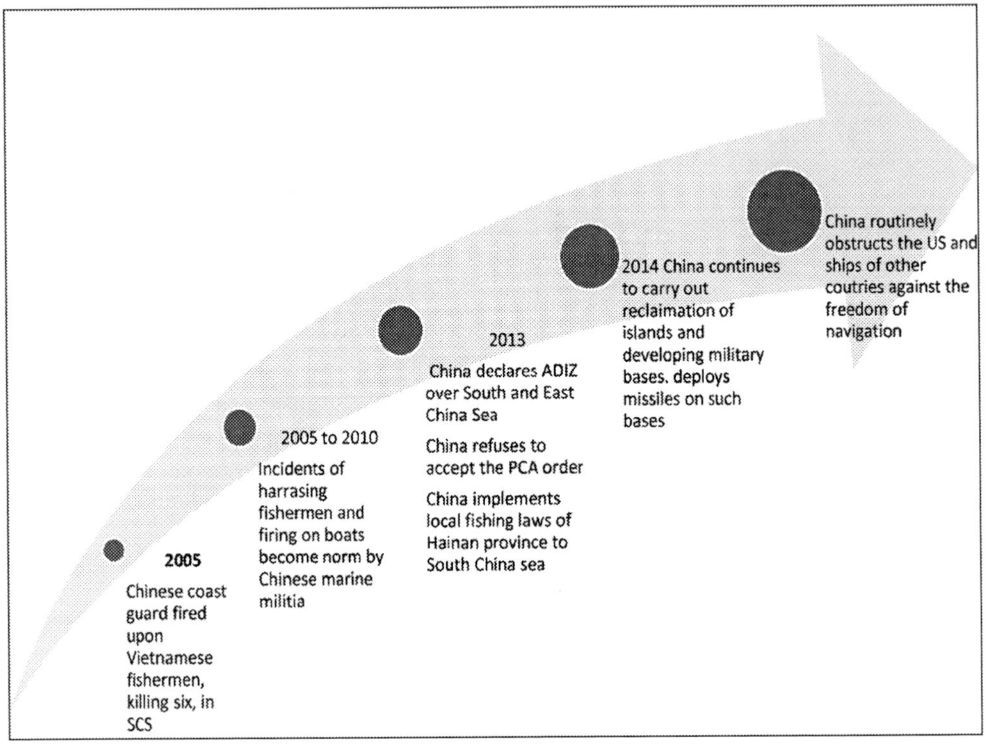

Source: Compiled from data given in the chapter

local laws in the South China Sea, use of media for creating a rhetoric and to sell the narrative are examples. The second element is the physical actions at the functional level of dispute like 'Cabbage strategy' or 'Salami slicing strategy' wherein physical presence at the site of contention in large numbers or create incidents repeatedly to establish the claim. The third element is the use of economic and diplomatic coercion. Case studies of Senkaku Island and Scarborough Shoal bring out how China brought down the resistance of Japan and the Philippines. The fourth element is the use of force wherever required to establish the claim and physical occupation of the disputed site by stealth and develop infrastructure at the fastest possible pace, so that the situation becomes a fait accompli for other stake holders. The Chinese have been very astute to keep the use of force below the threshold level. The reclamation and construction of seven such islands and completion of military bases including air strips in three of the islands is a case in point. The elements are explained graphically in Figure 4.12.

Figure 4.12: Elements of Chinese Hybrid Warfare in South and East China Seas

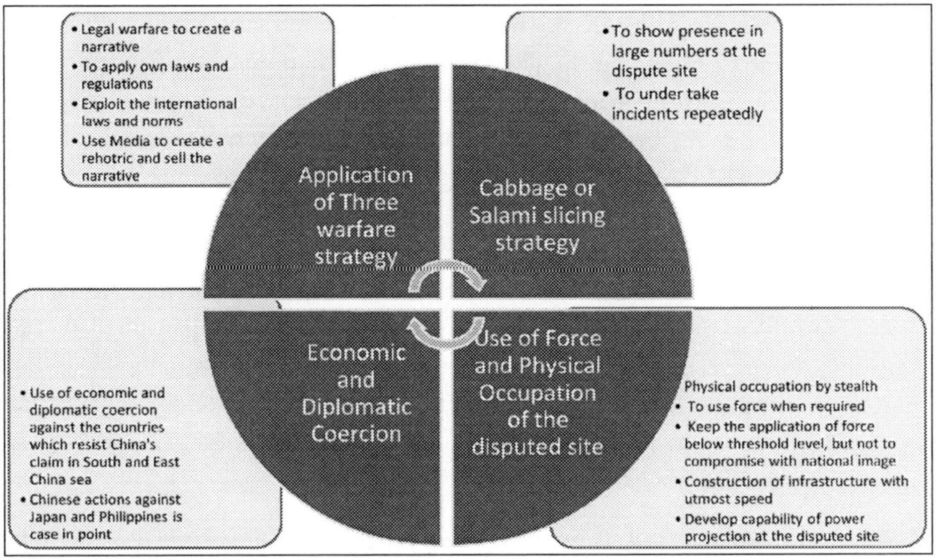

Source: Compiled from data given in the chapter

Elements of Chinese Hybrid Warfare in South and East China Seas

It is evident that China has been able to extend the influence and progress its claims in the South and East China seas by application of various means of hybrid warfare simultaneously and gradually. The Chinese have been able to achieve gains and also change the status unilaterally without crossing the red line. How China has used 'Economic warfare' as the most effective weapon of its 'Three-Warfare Strategy has been examined' in the next part of the chapter

Geo-economics: The Most Effective Weapon in Chinese Strategy of Unrestricted War

The extraordinary rise of China in the hierarchy of power is the direct outcome of spectacular economic growth. The 30 years of robust economic growth has provided China with the required economic and muscle power to assert itself in the comity of nations. China has successfully used its increased economic power in shaping the geopolitics to its advantage. The use of economics is inherent in the 'Three Warfare' strategy, whether it is influencing public opinion or fighting a legal war. This part of the chapter will bring out how China has used its economic might as the most potent weapon of its strategy of unrestricted warfare.

Preserving the Core Interest: Reunification with Taiwan. Reunification with Taiwan remains the core interest of China. On January 2, 2019, President Xi in his speech claimed that the Chinese dream (of national reunification) is the common dream of compatriots across the Strait.[83] Earlier, on the occasion of the 19th Congress on October 18, 2017, Xi had announced that national reunification would be achieved by 2049.[84] Not ruling out the use of force, albeit in the last few decades, China has put in place a full range of geo-economic instruments to preserve its core interests. Armed with economic strength, China now relies more and more on a strategy of economic encirclement and penetration to force Taiwan towards unification.[85] These instruments or policies are being applied in two forms; first being the multilateral, with the aim of encirclement, i.e., how China has been using geo-economics to isolate Taiwan, and secondly by bilateral engagement, in which how it uses economic policies to engage Taiwan directly, thereby penetrating Taiwanese society.[86]

Diplomatic Isolation of Taiwan. In accordance with 'One China Policy' China proclaims Taiwan as an inalienable part of China. Hence technically no country can have diplomatic relations with China and Taiwan simultaneously. China has been aggressive in pushing its One-China Policy and has been using economic threats in terms of sanctions against countries which endorse Taiwan's independent status.[87] In 2018 alone, El Salvador and Burkina Faso have denounced their diplomatic relations with Taiwan.[88] There are reports that when Taiwan showed its inability to fund construction of a port for El Salvador due to reasons of non-viability, the Chinese stepped in and the rest is history. Now, there are only 17 countries left that have diplomatic relations with Taiwan as an independent country.

Penetrating Taiwanese Society. China has been exerting economic pressure which impacts the life of the normal Taiwanese citizen. In 1995, when ROC President Lee Teng-hui gave a speech asserting the de facto status of ROC on Taiwan, China reacted with a threat of war by test firing missiles and conducting military exercises. Simultaneously, there was a crash in the Taiwan stock exchange (TAIEX) which led to a loss of approx. 30 percent of its total value. While China denied any direct hand, speculation was strong of a Chinese hand. [89] The Chinese ability to penetrate directly improved from the year 2000 onwards when the 50-year-old ban in Taiwan to trade and invest directly with mainland China was removed by the President of Taiwan. China became the largest trading partner of Taiwan in just three years of lifting of the ban. Today, China (including Hong Kong) account for approximately 43 percent of Taiwanese exports.[90] This growing dependence of Taiwan on China has helped the latter to toughen its stance on political and military negotiations. China has been actively supporting the Kuomintang Party in the internal elections of Taiwan, which is known as a pro unification party. A clear indication of this was the huge jump in the TAIEX in 2008 when Ma Ying-Jeou (Kuomintang Party) was elected President of Taiwan. Ma paid back the Chinese by signing a cross-Strait currency swap agreement in August 2012. This provides Beijing an effective leverage in Taiwan's economy. China has been slowly and steadily making overtures that will indirectly push Taiwan towards unification.

Protecting Territorial Sovereignty. China has used economic warfare as a strategic weapon on the sovereignty dispute with Japan over the Senkaku Island chain.[91] In 2010, after the incident of arrest of the captain of a Chinese fishing trawler when it collided with Japanese Coast Guard ships, China retaliated by placing a sanction on the export of rare earth minerals and metals to Japan. These minerals are crucial components for electrical equipment which Japan exports to the USA and European countries. This ban forced Japan to not only release the prisoner, but also establish a new dimension.[92]

Having suffered the ban, Japanese industrial giants like Hitachi and Toyota shifted their plants to China. Ban on rare earth materials needs to be seen as economic coercion by which China successfully used a crisis to achieve three strategic aims; first, to secure the release of the captain of the fishing trawler which had become an emotive issue; second to send a strong signal to Japan over its sovereignty rights on islands, and lastly to consolidate its market share in the global supply chain.[93] Later, as the Japanese government bought a portion of Senkaku Island, the dispute got aggravated; China became more aggressive in implementing trade sanctions and the ban on the export of rare earth minerals.

Expanding Influence in South East Asia. The geo-economic strategy of China in South East Asia apart from increasing its influence in the region aims to achieve three deeply interrelated objectives. These are firstly to check US alliances in the region, secondly to penalize and impose costs on any nation acting adversely on Chinese territorial and maritime claims and to preserve old friends like Cambodia, Laos and Myanmar.[94] Any opposition or representation against China's claim of the "nine dash line" in the South China Sea, on Paracel and Spratly islands, Scarborough and James shoals or many other such disputes extending up to Ntuma islands north of Indonesia is likely to invite an immediate reaction from China in terms of economic coercion.

In 2012, when a Philippine naval ship attempted to seize Chinese fishing trawlers close to the disputed Scarborough Shoal, China banned the import of 15 containers of bananas from the Philippines, followed by a ban on other fruits and vegetables which hit the local farmers hard. As the country was managing the woes of farmers, China swiftly occupied the shoal.[95] The vulnerabilities of ASEAN countries increases with the fact that China not only remains the largest trading partner of ASEAN as a block, but individually also it is the largest trading partner with most of the countries. The increased dependency of ASEAN nations on China is explained in the graph (Figure 4.13).

Figure 4.13: ASEAN–China Bilateral Trade

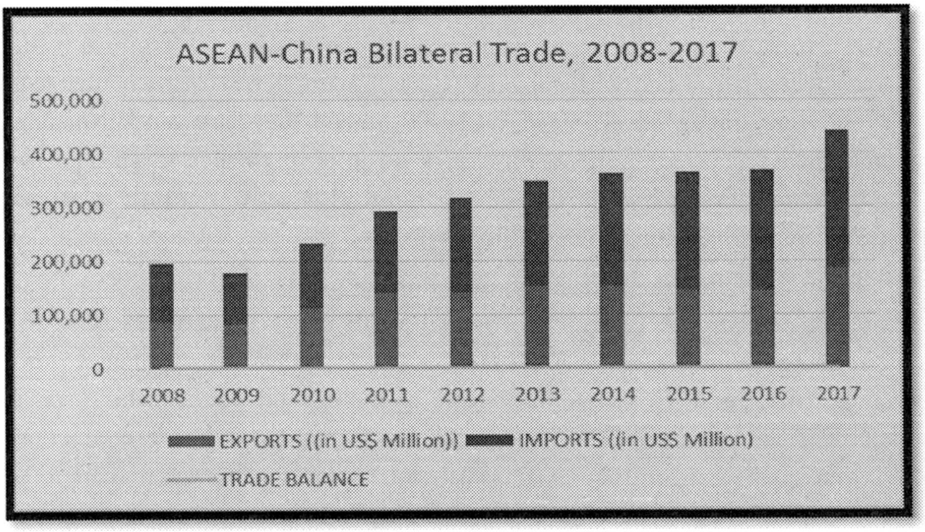

Protecting National Interest. There are numerous incidents that indicate that China has not hesitated in enforcing economic coercion against nations which have acted/apparently acted against the Chinese national interests. One the most sensitive issues is the interaction with or hosting the Dalai Lama. In a study of 2010, published in *Journal of International Economics,* which studied exports of 159 countries to China, it was found that countries whose leaders had the met Dalai Lama suffered a loss of trade with China in terms of exports falling from 8.1 to 16.9 percent.[96] In 2016, when Mongolia hosted the Dalai Lama, China reacted strongly by blocking the important trade route between the two nations and also imposing additional tariffs.[97]

In 2016-17, when the USA deployed THAAD missiles in South Korea, China hit back with economic sanctions against South Korea. As per a staff research report of the US-China Economic and Security Commission of 2017, Chinese regulators banned a considerable range of Korean products from being exported to China like cosmetics, air purifiers, video games, etc. By the end of the first quarter in 2017, the consumer export had dropped by 5.6 percent. Korean auto giants like Hyundai and Kia had a 52 percent drop in sales in China. The industry to suffer the biggest hit was tourism in Korea, as 47 percent of foreign tourists visiting Korea were Chinese. There was a drop of 66 percent in Chinese tourists to Korea.[98] In 2009, some Uighurs had taken shelter in Cambodia and were seeking asylum. However, in spite of UN objections, Cambodia repatriated them to China to face prosecution in connection with violent protests that had taken place in Xinjiang. It is reported that, Cambodia succumbed to the pressure and threat advanced by China. Cambodia soon was gifted with grants and loans worth 1.2 billion US dollars by China.[99]

BRI, the Great Chinese Dream or the Tool of Debt Trap Diplomacy. The ambitious Belt and Road Project is a manifestation of Chinese economic diplomacy, wherein China has been targeting vulnerable nations through loans and getting them to fell into a debt trap. President Xi, during the 19th National Party Congress, formally adopted the Belt and Road Initiative under the Party Constitution as part of a resolution to achieve "shared growth through discussion and collaboration." The project is being seen as the great Chinese Dream and an ambitious attempt to challenge US domination. While, these aspects are true, the more important fact is that China needed new markets to support the export-based economy. The economic growth has remained static since the early part of the second decade of the century. The Chinese White Paper on Defence of 2013, laid out a new role of the PLA to protect Chinese assets beyond the mainland. It was envisaged by the Chinese government that expansion of trade to new frontiers is a must for survival of the state. Hence, the manifestation of the BRI is an outcome of necessity and is being sold as "shared growth through discussion and collaboration" by the Chinese government.

The BRI program is a win-win situation for both the Chinese government and the Chinese infrastructure conglomerates. Through BRI, by constructing infrastructure in poor countries, China has been able to manage the excess availability/supply of dollars which in any case was not generating enough returns. China is offering loans to economically weak nations for infrastructure projects. The contract for such projects is being granted to Chinese companies which further employ Chinese labourers. Apart from generating employment, China has also been able to increase its strategic influence and capture new markets. China knowing very well that these countries will not be able to pay back the loan is investing in these countries. As a matter of fact, the terms of loans are made such that on the face of it the offer looks too attractive to refuse, but soon even the payment of interest on these loans becomes unviable. According to the Centre of Global Development policy paper of 21 March 2018, the contracts are opaque and there is total lack of transparency and contracts generally do not follow international norms.[100]

There have been cases wherein countries have been lured into construction of infrastructure projects which they did not even need. The construction of a port at Hambantota in Sri Lanka is one such case in which the Chinese offered to construct a port which did not have any economic viability. The offer made was such that the Sri Lankan government fell into the trap. On completion of the port, the Sri Lankan government was paying 90 percent of its revenue earned in servicing the debt it had taken for construction. In order to pay the accumulated interest and service the debt, the Sri Lankan government had to lease out the Hambantota port to Chinese for a period of 99 years.

Another case study is of Montenegro, in which the country accepted a Chinese proposal to construct a super highway connecting the port of Bar on the Adiatric Sea to landlocked Serbia. After the first phase of construction, Montenegro is under a debt of approximately one billion USD and for the remaining three phases approximately another 1.2 billion USD would be required. For Montenegro, a country of just 620,000 people, even the payment of interest in not feasible. As per a World Bank report, there is a minimum requirement of 25,000 vehicles using the highway to justify the construction However, as per current estimates, the traffic on the busiest stretch will not cross 6,000 vehicles.[101] The debt-to GDP-ratio in Montenegro has gone beyond 80 percent leading to an increase in taxes and a temporary freeze on public sector wages. The feasibility studies carried out by the French firm Louis Berger for the Montenegrin government, and the second by the U.S. company URS for the European Investment Bank have opined negatively on the prospect of the project. However, another study funded by the Export-Import Bank of China found the project viable, though the findings were never made public. A point of interest is that the first phase of the project was constructed by China Road and Bridge Corporation (CRBC) with loans from the Export-Import Bank of China.[102] This raises questions on the entire process and the business practices followed by China.

Encircling India with Infrastructure - Debt Trap Diplomacy. In Asia, China and India are the two biggest countries in terms of geography, economy, defence forces and human resources and also happen to share the longest disputed land border. Both the countries are also amongst the fastest growing nations. Hence, competition, rivalry and conflict in the common areas of influence are natural. While, India considers China as the biggest strategic challenge, China too aims to slow down or restrict the increasing influence of India. China has been able to achieve the strategic aim of encircling India and increase its strategic leverage in the Indian neighbourhood by successfully using its economic might.

The China-Pakistan Economic Corridor (CPEC) is the magnum opus of the Chinese infrastructure trap diplomacy. The CPEC connecting Gwadar to Karakoram Highway (KKH) has both economic and strategic importance for China and Pakistan. As per a *Nikkei Asian Review* report, China is the biggest lender to Pakistan accounting for more than 19 billion USD out of a total debt of approximately 90 billion USD.[103] The bulk of Chinese loans are due to the funding of the CPEC. The capacity of Pakistan to pay back the loan is under severe scepticism as, in 2109, Pakistan was on the brink of financial collapse and was rescued by Saudi Arabia and the UAE.[104] A point to note is that the payment of CPEC loans are yet to begin. As a collateral to the loan, Pakistan has handed over the operation of Gwadar port to China for a period of 40 years. It is reported that as per the agreement the state-run China Overseas Port Holding Company will take 91 percent of revenue collected from terminal and marine operations and 85 percent of gross operations revenue from the Gwadar free zone.[105] There are reports of China establishing a military base at Jiwani which is just 80 kilometres West of Gwadar.[106] The CPEC connecting Gwadar to KKH which further connects the Western Highway encircles India and provides China a strategic advantage.[107] Access to Gwadar Port, strategic highway all along India's frontier and a vassal state which has lost its strategic autonomy provides a tremendous strategic advantage to China against India in a long term-perspective.

Apart from Pakistan, the increasing Chinese footprints in Nepal, Myanmar, Bangladesh, Sri Lanka and Maldives are of concern to India. Chinese investments in these have been aggressive and not always in industry. In fact the major share of investments have been in infrastructure and energy. Together, these sectors claim 83 percent of Chinese investments in these countries. Nepal is the next country after Sri Lanka and Pakistan, which is on the verge of falling into the debt trap of China. China is the largest foreign investor in Nepal. Investment has shot up from just 1.81 million USD in 2003 to 174 million USD in 2014/15.[108] In 2017-18, Nepal received 550 million USD from China which is 84 percent of the total FDI received.[109] This shows a 500 percent hike in Chinese investments in Nepal, which has provided China the required leverage with Nepal to be used against India. Nepal pulling out of the joint military exercise BIMSTEC in 2018, on the insistence of China is one such manifestation. China has invested heavily in Myanmar and Bangladesh, however, these two countries have taken the lesson from Sri Lanka and have become conservative with Chinese investments. Still, China has been able to increase its leverage with these two countries substantially in the last few years.

Increasing Influence in IOR by Capturing Ports at Strategic Locations by Economic Coercion. The Chinese dependence on the sea lanes of communications in the IOR region is a known fact. More than 80 percent of Chinese trade passes through the IOR. China has been using all ways to increase its influence in the region. In the last few years, it has acquired/built/operated from strategically located ports in the IOR. The 'string of pearls' strategy has been discussed widely amongst scholars, academia and strategists. But what is more important is how has China gone about gaining control over these ports. The ports of Hambantota in Sri Lanka, as discussed in earlier paragraphs, have been acquired by aggressive and predatory economic takeover. Kuntan in Malaysia, KyaukPyu deep-water port in Myanmar, Hambantota in Sri Lanka and Gwadar in Pakistan are some of the ports that China is funding and building. Another country under heavy Chinese debt trap and may end up constructing and subsequently handing over a port to China is Maldives, China has shown a keen interest in constructing a port in Maldives.[110] According to Reuters and Forbes, Maldives already owes China 3.2 billion USD, which is more than 8,000 USD per individual of the small country.[111] The debt-to GDP-ratio of Maldives has crossed 100 percent. Maldives is now moving on the same path as Sri Lanka, which had to undergo a debt swap agreement and hand over the Hambantota port to China.

The Chinese strategy to increase their influence in South Asian countries on the periphery of India is by bringing in huge investments in Pakistan, Sri Lanka, Nepal, Bangladesh and Myanmar and this is a classic example of the use of geo-economics as a tool of first resort in aggressively pursuing foreign policy aims. Along with increasing its footprint on the periphery, China has been putting direct pressure on India on various issues like, asylum to Dalai Lama, claim on

Arunachal and the boundary issue. In the next part of the chapter, China's hybrid warfare against India and its future contours have been discussed.

China's Hybrid Warfare against India

China's hybrid war against India has not been very overt; rather, China has been exploiting the fault lines in India in a covert manner. It China exploited two basic weaknesses in Indian governance. First, was the failure of the Indian government to integrate the North Eastern region with the mainland. This left a bitter taste and a feel of alienation amongst the tribal communities of Northeast India. There is clear evidence of China providing moral, material and ideological support to the Naga, Manipur and Mizoram insurgents in the 1960s.[112] The second fault line was the failure of the Indian administration to ameliorate the plight of the rural, landless farmer class of the lower Ganga valley in the state of Bihar and West Bengal. The rural uprising of Naxalbari sought ideological motivation from the teachings of Mao.

Naxalite Movement. When "Spring Thunder" was launched in Naxalbari in 1967, the Chinese Communist Party mouthpiece wrote in its editorial in 1967, "*A peal of Spring Thunder has crashed over the land of India.*" The editorial signalled the moral support from the highest echelon of the CCP in China.[113] In 1967, a group of about 12 Naxalites travelled to China through Nepal, where they were provided ideological, political and weapons training.[114] On April 9,1970, Radio Beijing announced, "*at present, the flames of the peasants' armed struggle have spread to West Bengal, Bihar, Uttar Pradesh, Himachal, Orissa, Assam, Tripura and particularly Andhra Pradesh.*" (sic) The Chinese support to the Naxal movement could have been more dangerous, as Chinese had suggested that the CPI(M) should form regular armed forces.[115] The Chinese interest in CPI(M) reduced with Deng Xioping coming into power.

Legal Warfare. China claims that the entire Arunachal Pradesh is a part of Southern Tibet, based on twisted historical facts. In support of this narrative, in January 2007, China denied a visa to an Arunachal-Pradesh born IAS officer when, just a few months earlier, on 21 November 2006, a joint declaration had been signed between the two nations which reflected the shared vision and mutual objectives of the two countries on a number of important areas like: economic cooperation, international economic order and regional cooperation, civil nuclear cooperation, boundary questions and cultural exchanges.[116]

Economic Coercion. China has been interfering in the development of Arunachal Pradesh calling it disputed territory. In 2010, China blocked Asian Development Bank Funding for a construction of a dam in Arunachal Pradesh.[117] Strategic analysts observed that this incident could herald future conflicts once China gains influence in other international/ multilateral organisations like the IMF and starts interfering in future initiatives based on what Beijing sees as its interests.[118] Assam Chief Minister Tarun Gogoi had claimed in 2011 that the Maoists and the anti-talks faction of the ULFA were involved in the agitation against the construction of a hydro-electric project in Lakhimpur district.[119] The Chinese connection of Maoist groups and the anti-talks faction of the ULFA has already been established.

Cyber Warfare. Chinese cyber warfare capability has been discussed in detail earlier in the chapter. One of the salient aspects, rather the strength of cyber warfare, is the anonymity. It is near impossible to attribute or find the source responsible for the attack unless the source wants it to be known or conveys the message. Hence, it is difficult to pinpoint the blame of cyber-attacks on sensitive and critical Indian networks to China. According to a report called 'Shadows in the Clouds' released by Canadian and US cyber researchers in Toronto in 2010, a Chinese cyber spying operation called 'Shadow Network', attacked Government of India and Ministry of Defence sites and stole sensitive information on the Dalai Lama and the Indian security establishment.[120] Researchers, earlier in 2009, had reported that a network called 'Ghost Net' that was operating from servers in China had targeted official computers of the Government of India and stolen information related to the Dalai Lama's correspondence with India.[121] In the last few years, a number of suspected cyber-attacks on critical infrastructure have been reported. Against some of these, the Government has lodged FIRs and started investigations, but more often the government has not acknowledged the failure of the system due to cyber-attack. One such case is the shutting down of all systems of the T-3 terminal of IGI airport in New Delhi on 29 June 2011. After an investigation lasting three months the Government acknowledged that it had been caused by a "malicious code" sent from a remote location to breach the security of the airport. Accordingly, a case was registered with the CBI under IT and IPC.[122, 123] According to a report of NDTV, 4.3 lakh cyber-attacks were launched on India from January 2018 to June 2018, of which 42,544 originated in China.[124] In May 2015, the Minister of Communications and IT, Shri Ravi Shankar, informed Parliament that cyber-attacks were being launched from China and some other nations to penetrate computer networks of government organisations in India Details drawn from Parliament questions (PQ) regarding cyber-attacks/hacking of government sites for the last five years is given in Table 4.3. Figures quoted in the Parliamentary questions have been taken from CERT.

Table 4.3: Details of Cyber Attacks/Hacking of Web Sites

Sl.No.	*Year*	*Number of Cyber Attacks/ Hacking of Government sites*	*Reference of the PQ*	*Remarks*
1.	2014	44679	Starred Question No. 16, in Lok Sabha on 06.11.2016[125]	Cyber attacks
2.	2015	49455		
3.	2016	33147	Starred Question No. 31, in Lok Sabha on 12.12.2018[126]	Hacking of Indian web sites
4.	2017	30067		
5.	2018	15779 (till Nov 2018)		

Source: Figures taken from Parliament Questions

Chinese Assertiveness on Indo-China Border (LAC). The India–China boundary problem started with the Chinese annexation of Tibet in 1950. This was the first time that the geographical and political buffer preserved by the British Government was demolished, and this new-found contiguous border became one of the major drivers to govern relations between the two of the largest nations in the region. Former Foreign Secretary Kanwal Sibal maintains that it was the failure on India's part to visualise the impact of the Chinese annexation of Tibet in the absence of a formally demarcated boundary in Jammu and Kashmir and China not acknowledging the McMahon Line in the east and that the potential boundary problem should have been anticipated by Indian policy makers during that time.[127] The Chinese unilaterally changed the ground situation in Aksai Chin by constructing a road connecting Xinxiang province, which forced India to move security posts forward to prevent any further loss of territory, which ended being one of the triggers for the 1962 war. The war came as a shock to India as it had played the role of a friendly and supporting neighbour to China in the initial days, whether it was candidature to become a Permanent Member of the UN Security Council or recognising Tibet as an autonomous region of China. It even signed the 1954 Trade Agreement with the Tibet region of China, thereby de jure accepting Tibet as an integral part of China. After the 1962 war, there was another conflict at Nathu La in 1967. Since then, there has been no major incident involving direct or indirect combat on the India-China border. However, there have been incidents like Sumdorong Chu in 1987 and Doklam in 2017 which had reached at a point that could have triggered the second Indo–China war. The various CBMs between the two nations like "Agreement on Maintenance of Peace and Tranquillity along the Line of Actual Control-1993"; then in 1996;[128] and the Border Defence and Cooperation Agreement (BCDA) signed in 2013 have been largely effective. The situation on the border remains fragile and sensitive. As the border is not demarcated, each country has its own understanding of the LAC. Simultaneously, each country has its own perception of understanding of the LAC by the other side. Hence, a patrol coming face-to-face is not unusual. However, in this decade, such face offs between patrols have not only increased, but have escalated to stand offs, major incidents; the last one at Doklam being the longest, i.e., over 70 days. This pattern of increasing hostility during the face offs can be attributed to dynamic factors like perception of the LAC at the point of contact, sensitivity of the area where contact has taken place, personality of local commanders and also may be directions from higher ups. There is no official release of figures of violation of LAC by patrols of either side. Data based on figures of transgressions and major incidents as given out in newspapers and in response to Parliamentary Questions is given in Table 4.4.

Table 4.4: Details of Transgressions by PLA patrols across LAC and major incidents

Sl.No.	*Year*	*Chinese Transgressions*	*Dates of Major Incidents*	*Coinciding with Major Event*
1	2010	228[129]	–	Dec 10 visit of Premier Wen Jiabao; prior to that Beijing denied visa to Northern Army Commander
2	2011	213[130]		
3	2012	426[130]		
4	2013	411[132]	13 Apr, PLA patrol came 10 km inside the border in Eastern Ladakh (Depsang), and camped there for a few days	19 May, Premier Li Keqiang's state visit
5	2014	500[133]	• In Apr at Demchok • On Sep 14, PLA troops intruded in Chumar, stayed for more than a week.	President Xi visited India from 19 to 19 Sep. The Chumar incident was during President Xi's visit.
6	2015	350[134]		
7	2016	273[135]	• In Mar at Demchok • Yangtse, in Arunachal Pradesh	
8	2017	426[136]	• On Aug 18, at Pangong Tso, face off leading to stone pelting incident. • Doklam (it was not transgression)	
9	2018	170 (Aug 2018)[137]		

Source: Figures taken from Parliament Questions and newspaper reports

The number transgressions as given in the table indicate that transgressions by Chinese patrols have increased significantly since 2012; it came down in 2015 and then went up again in 2016. Complete data for 2018 is not yet available. On analysing the major incidents, apart from the pattern of transgression, something significant stands out, i.e., the behaviour of PLA patrols when there is an important visit or event. In April–May 2013, just before the visit of Li Keqiang on 19 May 2013, there was a major intrusion in the Depsang Plains. In 2014, during the landmark visit of the Chinese President, there was a faceoff at Chumar. A similar pattern was observed after the stand-down at Doklam. The Whuhan summit was held in April 2018, the next major event or interaction between Prime Minister Modi and President Xi was in June 2018 on the side lines of the SCO Summit. In the intervening period of three months, there were 28 transgressions by the PLA.[138] Such intensity of patrolling can only be construed as political signalling as was done in 2013 and 2014.

China has also been changing the status of the LAC unilaterally by a creeping-forward technique or 'salami slicing' as discussed earlier. In January 2018, the Chinese attempted to construct a track on the Indian side of the LAC near Bishing village in Tuting.[139] On being stopped by Indian troops, the Chinese went back and their road-construction equipment was seized by the Indian Army. What is significant about the incident is that though the site is on the LAC no faceoff, transgression or any kind of patrolling activity by the PLA had been reported earlier. This shows that the Chinese keep creating new areas of dispute on the LAC to keep it active. Earlier, the Chinese had constructed a road in Pangong Tso area in Eastern Ladakh, which is 5 km inside the Indian perception of the LAC.[140] It is a well-established modus operandi of the PLA. First, its patrols probe the LAC; if there is no negative reaction, then they repeatedly visit the area and subsequently they occupy it. It is quite similar to what the Chinese strategy had been in the South and East China Seas.

Another aspect that merits attention is China's management of land borders since 1949. It has 12 land and six maritime neighbours: and with Vietnam and North Korea, it shares both land and maritime borders. It is a matter of interest to note that of 14 countries with whom China had land border disputes, it has been able to reach mutual settlements with 12; only India and Bhutan are pending.[141] Professor M. Taylor Fravel, in his very well researched article in the International Security paper of Harvard College, has carried out a detailed analysis on Chinese border issues and China's attitude in resolving border issues.[142] An extract of China's boundary issues is given in Table 4.5.

Table 4.5: China's Boundary Issues

Sl.No.	*Country*	*Area (Km²)*	*Year of Settlement*	*Description of Settlement*
1	Myanmar	1909	1960-61	China received 18% of disputed land, Myanmar received the strategic highway, salt mine and preferred watershed boundary in North (McMahon line)
2	Nepal	2476	1960-63	China received 6% of disputed land and half of Mt Everest, Nepal held most of the grazing areas and passes
3	North Korea	1,165	1962	China received 40% of disputed land, North Korea received the balance
4	Mongolia	16,808	1962-64	China received 29% of the disputed land
5	Pakistan	8806	1963-65	China received 60% of disputed land but transferred 1942 sq km to Pakistan
6	Afghanistan	7381	1963	China did not receive any part of the land in the disputed Wakhan corridor
7	Russia • Eastern part • Western part • Strategic islands	 • 1000 • No data • 408	 • 1999 • 1999 • 2004	 • China received 52% of river islands, balance distributed equally • Agreement affirmed actual line of control • Control of Abagaitu and Hexiazi islands were divided evenly
8	Laos	18	1991-93	China reportedly received 50% of disputed land
9	Vietnam	227	1999	China received 50% of disputed land
10	Kazakhstan	2420	1998-02	China received approx. 22% of disputed land
11	Kyrgyzstan	3656	1999	China received approx. 32% of disputed land
12	Tajikistan	28430	2002	China received 4% of disputed Pamirs, balance land was evenly shared
13	Bhutan	1128	On-going	Final settlement yet to be arrived at
14	India	1,25,000	On-going	Final settlement yet to be arrived at

Source: The inputs have been taken from the paper of Professor M. Taylor Fravel[143]

It is evident from the table above that of the 12 countries with which China has resolved boundary issues, 10 of them have been on compromise. In the case of Myanmar, China has accepted the McMahon Line and watershed principle, but has refused to accept the same with India. On one hand, it is willing to resolve boundary issues with other countries on the principle of compromise, whereas with India it is has been increasing its claim. In 2006, on the eve of the visit of President Hu Jintao, it asserted its claim over entire Arunachal Pradesh.[144] Subsequently in 2007, it refused a visa to an IAS officer who was from Arunachal Pradesh on grounds that the state is a part of China.[145] Later, in 2009, it stalled the Asian Developed Bank Loan calling Arunachal disputed territory.[146] Why is China not willing to resolve the boundary issue with India? The answer is, that it does not want to. China understands the cost of guarding the LAC for India, the repeated territorial incursions or transgressions like Dpesang, Chumar and Doklam, keeps India off balance politically and militarily. It imposes a huge economic cost on a developing country like India, which is its potential rival in the regional power calculus. China will continue to leverage the unsettled borders to its advantage. It is a classic case of hybrid warfare as it meets more than four parameters of the framework that have been stated in Chapter 1. It is prolonged and has lasted for more than 50 years. It remains sub-conventional and enforces economic and political cost to the adversary. The coercive incidents so far have been restricted to physical altercation but nowadays not crossed the threshold of small arms fire. However, during such incidents, emotions run high and there is a high possibility of an incident that can lead to a border skirmish and flare up into a violent conflict.

Conclusion

China is home to Sun Tzu, and has one of the longest traditions of strategic thinking. Sun Tzu, in '*Art of* War', treats war as a process of diplomacy; in essence, war fighting is diplomacy by other means.[147] The Chinese have mastered the art of converting disputes into conflicts and resolving them to their advantage without crossing the threshold of war. China has successfully used a combination of economic appeasement, legal and psychological warfare along with coercion by military power to achieve its national aims. In the last decade, the Chinese have used economic might as one of the most effective tools to achieve the national will. The ambitious Chinese dream of BRI being sold as "shared growth through discussion and collaboration" is a manifestation of Chinese strategy to achieve the national will without fighting. The world is now taking cognizance of the '*Shi*' factor in the great Chinese dream of BRI and countries like, Malaysia, Bangladesh and Myanmar have now shown reservations in accepting the Chinese bait of loans and have rejected or downscaled a few BRI projects.

What will be the implications of the increasing Chinese economic engagements with Pakistan? How will CPEC affect the dynamics of POK? Will the China-Pakistan growing relationship remain purely economic or will it have military dimensions too? Is India prepared to face the increasing collusion between China and Pakistan? These are some of the questions which will be addressed in Chapter 5.

NOTES

1. Sun Tzu, The Art of War, translated by S.B. Griffith, Oxford: Clarendon press, 1964
2. Qiao Liang and Wang Xiangsui Unrestricted Warfare
3. Study for Andy Marshall, Director Net Assessment, Office of Secretary of Defence, USA, available at https://cryptome.org/2014/06/prc-three-wars.pdf accessed on 13 Mar 2019
4. Commodore Uday Bhaskar in China's Three Warfare concept related to India and in Indian Ocean in Study for Andy Marshall, Director Net Assessment, Office of Secretary of Defence, USA, available at https://cryptome.org/2014/06/prc-three-wars.pdf accessed on 13 Mar 2019
5. Professor Huiyun Feng, 'Chinese Strategic Culture and Foreign Policy Decision Making: Confucianism, Leadership, and War', Asian Security Studies, Routledge 2007, p. 02
6. Professor Huiyun Feng, 'Chinese Strategic Culture and Foreign Policy Decision Making: Confucianism, Leadership, and War', Asian Security Studies, Routledge 2007, p. 26
7. China's White Paper: China's Peaceful Development Road published by the State Council Information Office of Chinain 2005, available at http://unpan1.un.org/intradoc/groups/public/documents/ APCITY/UNPAN023152.pdf accessed on 20 Jun 2019
8. David Blasko in 'Marine Time Missions Require a Change in PLA Mindset, in *China Brief*, Vol. XV, Issue 12, June 19, 2015
9. Ibid
10. PLA White Paper 'China's National Defence in 2006' available at http://www.china.org.cn/english/features/book/194421.htm accessed on 31 Jan 2019
11. The Chinese White Paper of 2013, available at 'The Diversified Employment of China's Armed Forces' http://eng.mod.gov.cn/Database/WhitePapers/2013-04/16/content_4442756.htm accessed on 31 Jan 2019
12. Denis Blasko, in China Defence White Paper 2013, in James Town *China Brief*, volume 13, issue 9, available at https://jamestown.org/program/the-2013-defense-white-paper-in-perspective/ accessed on 2 Feb 19
13. Monika Chansoria in China's 2013 Defence White Paper: Charting "Diversified Employment of Armed Forces" available at www.claws.in/images/journals_doc/06-China%C6s%202013 %20 Defence%20White% 20Paper. pdf, accessed on 3 Feb 2019
14. Timothy R. Heath, in The 'Holistic Security Concept': The Securitization of Policy and Increasing Risk of Militarized Crisis, in *China Brief*, in James Town *China Brief*, volume 13, issue 9, available at https://jamestown.org/program/the-2013-defense-white-paper-in-perspective/ accessed on 2 Feb 2019
15. Ibid
16. Denis Blasko, in China Defence White Paper 2013, in James Town *China Brief*, volume 13, issue 9, available at https://jamestown.org/program/the-2013-defense-white-paper-in-perspective/ accessed on 2 Feb 2019
17. Timothy R. Heath, in The 'Holistic Security Concept': The Securitization of Policy and Increasing Risk of Militarized Crisis, in *China Brief*, in James Town *China Brief*, volume 13, issue 9, available at https://jamestown.org/program/the-2013-defense-white-paper-in-perspective/ accessed on 2 Feb 2019
18. Zhao Zijin and Zhao Jinfang, "On the Control and Management of Military Crisis," *China Military Science*, July 2013, pp 62-71, quoted by Timothy R. Heath, in The 'Holistic Security Concept': The Securitization of Policy and Increasing Risk of Militarized Crisis, in *China Brief*, in James Town *China Brief*, volume 13, issue 9, available at https://jamestown.org/

program/the-2013-defense-white-paper-in-perspective/ accessed on 2 Feb 19

19. Timothy R. Heath, in The 'Holistic Security Concept': The Securitization of Policy and Increasing Risk of Militarized Crisis, in *China Brief*, in James Town *China Brief*, volume 13, issue 9, available at https://jamestown.org/program/the-2013-defense-white-paper-in-perspective/ accessed on 2 Feb 19
20. Lt. Gen. Gautam Banjerjee, in *Consecration of China's 'New Period' People's Liberation Army*, published in *Strategic Discourse on The People's Republic of China*, Vikekananda International Foundation, Pentagon Press, pp, 175-193
21. Lt. Gen. S.L. Narsimhan, in 'PLA Reforms under Xi and their Security Implications' published in *National Security*, Vol 1, issue 2, Vivekananda International Foundation, available at https:/ /www.vifindia.org/sites/default/files/nov-2018-national-security-vol-1-issue-2-Narsihman.pdf accessed on 4 Feb 2019
22. David Blasko in PLA as Organisation V 2.0, in *PLA Ground Forces*, p. 262
23. PLA White Paper 'China's National Defence in 2006' available at http://www.china.org.cn/english/features/book/194421.htm accessed on 31 Jan 2019
24. David Blasko in PLA as Organisation V 2.0, in *PLA Ground Forces*, p. 262
25. Lt. Gen. Gautam Banjerjee, in Consecration of China's 'New Period' People's Liberation Army, published in *Strategic Discourse on The People's Republic of China*, Vivekananda International Foundation, Pentagon Press, p. 175-193
26. David Blasko in PLA as Organisation V 2.0, in *PLA Ground Forces*, p. 262
27. Lt. Gen. S.L. Narsimhan, in 'PLA Reforms under Xi and their Security Implications' published in *National Security*, vol 1, issue 2, Vivekananda International Foundation, available at https:/ /www.vifindia.org/sites/default/files/nov-2018-national-security-vol-1-issue-2-Narsihman.pdf accessed on 4 Feb 2019
28. David Blasko in PLA as Organisation V 2.0, in *PLA Ground Forces*, p. 262
29. Jhon Costello and Joe McReynolds in China's Strategic Support Force, in *China Strategic Perspectives*, no. 13 published by Centre for the Study of Chinese Military Affairs Institute for National Strategic Studies, National Defence University Press, Washington
30. John Costello and Joe McReynolds in China's Strategic Support Force, in *China Strategic Perspectives*, no 13 published by Centre for the Study of Chinese Military Affairs, Institute for National Strategic Studies, National Defence University Press, Washington
31. Ibid
32. Ibid
33. Ibid
34. Ibid
35. Ibid
36. Lt Gen SL Narshimhan, in PLA Reforms under XI and their Security Implications, in National Security, Volume I issue II November 2018, can be accessed at https://www.vifindia.org/sites/default/files/nov-18-national-security-vol-1-issue-2-Narsimhan-pdf
37. Jhon Costello and Joe McReynolds in China's Strategic Support Force, in China Strategic Perspectives, No 13 published by Center for the Study of Chinese Military Affairs Institute for National Strategic Studies, National Defense University Press, Washington, p.11
38. John Costello and Joe McReynolds in China's Strategic Support Force, in *China Strategic Perspectives*, no 13 published by Centre for the Study of Chinese Military Affairs, Institute for National Strategic Studies, National Defence University Press, Washington, p.11

39. Indo-Pacific Strategy Report, The Department of Defence. 1 June 2019, p. 8
40. *Indo-Pacific Strategy Report,* The Department of Defence. 1 June 2019, p. 8
41. Abhay K. Singh, *Expanding the Turbulent Maritime Periphery, Hybrid Warfare The Changing Character of Conflict*, IDSA, Pentagon Press, p. 149
42. Ibid
43. *South China Morning Post,* available at https://www.scmp.com/week-asia/explained/article/2187161/explained-diaoyu/senkaku-islands-dispute accessed on 1 June 2019
44. Diayou Islands Dispute: Perspective from China, *The Diplomat,* Aug 8, 2018, https://thediplomat.com/2018/08/diaoyu-islands-dispute-a-chinese-perspective/accessed on 1 June 2019
45. *South China Morning Post,* available at https://www.scmp.com/week-asia/explained/article/2187161/explained-diaoyu/senkaku-islands-dispute accessed on 0 June 2019
46. Senkaku Islands Dispute, *South China Morning Post* available at https://www.scmp.com/week-asia/explained/article/2187161/explained-diaoyu/senkaku-islands-dispute accessed on 2 June 2019
47. Abhay K. Singh, *Expanding the Turbulent Maritime Periphery, Hybrid Warfare The Changing Character of Conflict*, IDSA, Pentagon Press, p. 150
48. Trends in Chinese Government and Other Vessels in the Waters Surrounding the Senkaku Islands, and Japan's Response, Ministry of Foreign Affairs, Japan available at https://www.mofa.go.jp/region/page23e_000021.html accessed on 3 June 2019
49. Arrest in Disputed Waters, *New York Times,* Sep 19, 2010 available at https://www.nytimes.com/2010/09/20/world/asia/20chinajapan.html accessed on 2 June 2019.
50. For more details see Jeff Himmelman, *Game of Shark and Minnow, The New York Times* available at http://www.nytimes.com/newsgraphics/2013/10/27/south-china-sea/index.html accessed on 5 June 2019
51. Trends in Chinese Government and Other Vessels in the Waters Surrounding the Senkaku Islands, and Japan's Response, Ministry of Foreign Affair, Japan available at https://www.mofa.go.jp/region/page23e_000021.html accessed on 3 June 2019
52. Explained: Diaoyu/Senkaku islands dispute, *South China Morning Post,* available at https://www.scmp.com/week-asia/explained/article/2187161/explained-diaoyu/senkaku-islands-dispute, accessed on June 2019
53. For more, see Japan Scrambles Fighter Jets 999 Times in 2018 in Response to Foreign Aircraft, *The Diplomat,* available at https://thediplomat.com/2019/05/japan-scrambles-fighter-jets-999-times-in-2018-in-response-to-foreign-aircraft/accessed on 4 June 2019
54. Chinese Activities Surrounding Japanese air space, https://www.mod.go.jp/e/d_act/ryouku/ ,accessed on 3 June 2019
55. Abhay K. Singh, *Expanding the Turbulent Maritime Periphery, Hybrid Warfare The Changing Character of Conflict*, IDSA, Pentagon Press, p. 151
56. For more, see 'How Much Trade Transits the South China Sea? Not $5.3 Trillion a Year', *The Diplomat,* available at https://thediplomat.com/2017/08/how-much-trade-transits-the-south-china-sea-not-5-3-trillion-a-year, accessed on 7 June 2019
57. For more see, China's Maritime Dispute, Council on Foreign Relations, available at https://www.cfr.org/interactives/chinas-maritime-disputes?cid=otr-marketing_use-china_sea_Info Guide#!/chinas-maritime-disputes?cid=otr-marketing_use-china_sea_InfoGuide, accessed on June 6, 2019

58. For more, see 'How Much Trade Transits the South China Sea? Not $5.3 Trillion a Year', *The Diplomat*, Aug 7, 2017 available at https://thediplomat.com/2017/08/how-much-trade-transits-the-south-china-sea-not-5-3-trillion-a-year, accessed on 7 June 2019
59. For more details on historical background to the dispute in the South China Sea see, Congressional Research Service (CRS) report on 'China's Actions in South and East China Seas: Implications for U.S. Interests—Background and Issues for Congress', p. 11
60. O'Rourke, Ronald, in Congressional Research Service (CRS) report on 'China's Actions in South and East China Seas: Implications for U.S. Interests—Background and Issues for Congress, 'p. 18
61. Lyle J. Morris, in 'Time to Speak About South China Sea', Rand Blog, available at https://www.rand.org/blog/2019/03/time-to-speak-up-about-the-south-china-sea.html, accessed on 10 June 2019
62. Lieu Zhen in What's China's 'nine-dash line' and why has it created so much tension in the South China Sea?, *South China Morning Post*, available at https://www.scmp.com/news/china/diplomacy-defence/article/1988596/whats-chinas-nine-dash-line-and-why-has-it-created-so, accessed on June 8, 2019
63. Ibid
64. For more details, see CFR report on 'The Hague Tribunal's South China Sea Ruling: Empty Provocation or Slow-Burning Influence?' available at https://www.cfr.org/councilofcouncils/global_memos/p38227, accessed on 6 June 2019
65. Beijing Rejects the PCAs Ruling on South China Sea, *The Guardian*, available at https://www.theguardian.com/world/2016/jul/12/philippines-wins-south-china-sea-case-against-china, accessed on 6 June 2019
66. Abhay K. Singh, *Expanding the Turbulent Maritime Periphery, Hybrid Warfare The Changing Character of Conflict*, IDSA, Pentagon Press, p. 154
67. For more details, see O'Rourke, Ronald, in Congressional Research Service (CRS) report on 'China's Actions in South and East China Seas: Implications for U.S. Interests—Background and Issues for Congress,' p. 19
68. Derek Watkins in 'What China Has Been Building in the South China Sea', *The New York Times*, Oct 27, 2015, available at https://www.nytimes.com/interactive/2015/07/30/world/asia/what-china-has-been-building-in-the-south-china-sea.html, accessed on 6 June 2019
69. Ibid
70. For more details, see O'Rourke, Ronald, in Congressional Research Service (CRS) report on 'China's Actions in South and East China Seas: Implications for U.S. Interests—Background and Issues for Congress,' p. 19
71. For more details, see O'Rourke, Ronald in Congressional Research Service (CRS) report on 'China's Actions in South and East China Seas: Implications for U.S. Interests—Background and Issues for Congress,' p. 19
72. Ibid
73. Ibid
74. Philippine Warship Standoff with Chinese Fishermen, *The Guardian*, April 11, 2012 available at https://www.theguardian.com/world/2012/apr/11/philippines-china-stand-off-south-china-sea, accessed on 8 June 2019
75. Controversial Chinese oil rig drills first well in east South China Sea, *The Week*, available at https://www.theweek.in/news/world/2019/04/11/Controversial-Chinese-oil-rig-drills-first-

well-in-east-South-China-Sea.html

76. Vietnamese Fishermen versus China, in *New Mandala*, available at https://www.newmandala.org/vietnamese-fishermen-versus-china/, accessed on 9 June 2019
77. Ibid
78. China detains Vietnamese fishermen in disputed water, Reuters, March 22, 2012, available at https://www.reuters.com/article/china-vietnam/china-detains-vietnamese-fishermen-in-disputed-water-idUSL3E8EM3YJ20120322, accessed on 9 June 2019
79. China detains Vietnamese fishermen in South China Sea, *BBC News*, July 4, 2014, available at 2019, https://www.bbc.com/news/world-asia-28168447, accessed on June 9
80. Philippine fishermen accuse China of firing on vessel, *Indian Express*, April 21, 2017, available at https://indianexpress.com/article/world/philippine-fishermen-accuse-china-of-firing-on-vessel-4621963/, accessed on June 9, 2019
81. Chinese boats attack Vietnamese fishermen in South China Sea, UPI, available at https://www.upi.com/Top_News/World-News/2017/06/29/Chinese-boats-attack-Vietnamese-fishermen-in-South-China-Sea/8561498755312/, accessed on June 9, 2019
82. Vietnam protests to China over South China Sea boat sinking, *Reuters*, March 22, 2019 available at https://www.reuters.com/article/us-vietnam-southchinasea/vietnam-protests-to-china-over-south-china-sea-boat-sinking-idUSKCN1R307O, accessed on 9 June 2019
83. Keoni Everington in Chinese General lists 10 'privileges' Taiwan will 'enjoy' under 'one country, two systems' in Taiwan Times, Mar 5, 2019 available at https://www.taiwannews.com.tw/en/news/3651302 accessed on 05 mar 2019
84. Given in the script of the speech delivered by President Xi during the 19 Party Congress in Oct 17.
85. Roberet D Blackwill & Jenifer M Harris in War By Other Means, in Chapter on Geo-economics in Chinese Foreign Policy, p. 95, Harvard University press, London
86. Ibid
87. Ibid
88. Christ Horton, in El Salvador Recognizes China in Blow to Taiwan, New York Time available on https://www.nytimes.com/2018/08/21/world/asia/taiwan-el-salvador-diplomatic-ties.html
89. Robert D. Blackwill & Jenifer M. Harris in War By Other Means, in Chapter on Geo-economics in *Chinese Foreign Policy*, p. 95, Harvard University Press, London
90. The data taken from World's richest countries targeted research on Taiwan Exports, available at http://www.worldsrichestcountries.com/top_taiwan_exports.html, accessed on 5 March 2019
91. Brahma Chellaney in China's Weaponisation of Trade in *The Japan Times* available at https://www.japantimes.co.jp/opinion/2017/07, accessed on 5 March 2019
92. Roberet D. Blackwill & Jenifer M Harris in War By Other Means, in Chapter on Geo-economics in Chinese Foreign Policy, p. 95, Harvard University Press, London
93. Ibid
94. Ibid
95. Brahma Chellaney in China's Weaponisation of Trade, in *The Japan Times*, available at https://www.japantimes.co.jp/opinion/2017/07, accessed on 8 March 2019
96. Andreas Fuchs and Nils-Hendrik Klann in Paying a visit: The Dalai Lama effect on international Trade, published in *Journal of International Economics*, available on https://www.parisschoolofeconomics.eu/docs/koenig-pamina/article_dalailama_effect_on_internationaltrade_jie_2013.pdf, accessed on 8 March 19

97. China says it hopes Mongolia learned a lesson after the Dalai Lama's visit, *Reuters,* Jan 24, 2107 available at https://in.reuters.com/article/us-china-mongolia-dalailama-idINKBN158197, accessed on 8 March 2019
98. Staff research report of US-China Economic and Security Commission of 2017, available at https://www.uscc.gov/sites/default/files/Research/Report_China%27s%20Response%20to%20THAAD%20Deployment%20and%20its%20Implications.pdf, accessed on 8 March 2019
99. Robert D. Blackwill & Jenifer M. Harris in War By Other Means, in chapter on *Geo-economics in Chinese Foreign Policy*, p. 95, published by Harvard University Press, London
100. Priyanka Kher and Trang Tran in their discussion paper 'Investment Protection Along Belt and Road' published by Macroeconomics, Trade and Investment (MTI) Global Practice for World Bank, p. 30
101. Noah Barkin and Aleksandar Vasovic in Chinese 'highway to nowhere' haunts Montenegro, *Reuters,* available at https://www.reuters.com/article/us-china-silkroad-europe-montenegro-insi/chinese-highway-to-nowhere-haunts-montenegro-idUSKBN1K60QX, accessed on 11 March 19
102. Ibid
103. China Wrestles with Chinese Corridor Debt in *Nikkei Asia Review* available at https://asia.nikkei.com/Politics-Economy/International-Relations/Pakistan-wrestles-with-growing-Chinese-corridor-debt, accessed on 12 March 2019
104. Pakistan secures 6 billion USD aid package from Saudi Arabia, in Bloomberg available at https://www.bloomberg.com/news/articles/2018-10-23/pakistan-says-saudi-arabia-agrees-to-6-billion-support-package accessed on 12 March 2019
105. The China-Maldives Connection, in *The Diplomat* available at https://thediplomat.com/2018/01/the-china-maldives-connection/ accessed on 13 March 19
106. Rajeshwari Rajagopalan in A New China Military Base in Pakistan? Published in *The Diplomat* available at https://thediplomat.com/2018/02/a-new-china-military-base-in-pakistan, accessed on 12 March 2019
107. Sushant Sarin, monograph on *Corridor Calculus CPEC and China's Comprador Investment Model in Pakistan*, published by Vivekananda International Foundation
108. Prakash Bhandari in Growing Chinese Presence in Nepal, in *Asia Dialogue* available at http://theasiadialogue.com/2018/09/25/the-growing-chinese-presence-in-nepal, accessed on 12 March 2019
109. China tops Nepal's foreign direct investment list for third consecutive year, in *Global Times*, available at https://gbtimes.com/china-tops-nepals-foreign-direct-investment-list-for-third-consecutive-year, accessed on 12 March 2109
110. Maldives admits China interest in building port, in *Maldives Times*, available at https://maldivestimes.com/maldives-admits-china-interest-in-building-port-avas-mv, accessed on 13 March 2019
111. Panos Mourdoukoutas in "Modi Should Not Bail out China in Maldives", in *Forbes* available at https://www.forbes.com/sites/panosmourdoukoutas/2018/11/28/modi-should-not-bail-out-china-in-maldives/#24784a9a41d8, accessed on 12 March 2019
112. Understanding Indo-China conflict through history to contemporary times, Epilogue, available at https://epilogue.in/2017/07/28/understanding-indo-china-conflict-through-history-to-contemporary-times, accessed on June 10, 2019
113. R.S.N. Singh, in Is China Waging a Proxy War through the Maoists? *CLAWS Journal*, available

at https//www.archive.claws.in accessed on June 11, 2019
114. Bertil Lintner, in *Great Game East,* Harper Collins Publishers India p. 91, p. 342
115. Bertil Lintner, in *Great Game East,* Harper Collins Publishers India p. 91, p. 343
116. Ibid
117. China blocks ADB India loan plan, in *Financial Times,* web edition, 11 April 2009, available at https://www.ft.com/content/033935c2-25e4-11de-be57-00144feabdc0, accessed on 29 June 2019
118. Ibid
119. Maoists, ULFA Involved in Anti-Dam Stir: Gogoi, *Outlook,* e-pub, Dec. 28, 2011, available at https://www.outlookindia.com/newswire/story/maoists-ulfa-involved-in-anti-dam-stir-gogoi/745791, accessed on 28 June 2019
120. A report called '*Shadows in the Clouds*' prepared by Canadian and US researchers working in this field in Toronto claimed that a network operating from China called 'Shadow Network' through a systematic cyber espionage network may have stolen eight months of sensitive information on government, security, academics and on the Dalai lama. The network also targeted computers of Indian embassies and high commissions in other countries. For detailed report, see 'Chinese cyber agents hack in to India's secret documents' *Times of India* dated April 6, 2010, available at https://timesofindia.indiatimes.com/india/Chinese-agents-hack-into-Indias-secret-documents-Report/articleshow/5766129.cms, accessed on 25 June 2019
121. 'Spy Network Pilfered Classified Docs From Indian Government and Others', in '*Wire*' dated April 04, 2010, available at https://www.wired.com/2010/04/shadow-network/, accessed on 27 June 2019
122. 'Cyber-attack led to IGI shut down', in *Indian Express* web edition of 11 Sep. 2011, available at http://archive.indianexpress.com/news/cyber-attack-led-to-igi-shutdown/851365/,accessed on 29 June 2019
123. India is facing cyber-attacks from China, other nations: Ravi Shankar Prasad, *India Today,*web edition, dated May 7, 2015, available at https://www.indiatoday.in/technology/news/story/india-is-facing-cyber-attacks-from-china-other-nations-ravi-shankar-prasad-251835-2015-05-07, accessed on 30 June 2019
124. India Saw 4.3 Lakh Cyber Attacks from Nations Including US, China: Report, *NDTV,* Nov. 11, 2018, available at https://www.ndtv.com/india-news/india-saw-4-3-lakh-cyber-attacks-from-nations-including-us-china-report-1945695, accessed on 29 June 2019.
125. For more information on questions and answers, see Starred Question No. 16, in Lok Sabha on 06.11.2016, available at http://164.100.47.194/Loksabha/Questions/QResult15.aspx?qref=40475&lsno=16, accessed on June 29, 2019
126. For more information on questions and answers, see Starred Question No. 31, in Lok Sabha on 12.12.2018, available at http://164.100.47.194/loksabha/Questions/QResult15.aspx?qref=73726&lsno=16, accessed on 29 June 2019
127. Kanwal Sibal on India-China Problem and Prospects, in *Strategic Discourse on the People's Republic of China: Military Power and Politics,* Vivekananda International Foundation, 2017, Pentagon Press, p. 2.
128. Swaran Singh, in Sino-Indian CBMs: Problems and Prospects, IDSA Commentary, available at https://www.idsa-india.org/an-jul-4.html, accessed on 26 June 2019
129. No Chinese intrusion since 2010, only 'transgressions': Govt, in *Times of India* web edition, Aug. 20, 2014, available at https://timesofindia.indiatimes.com/india/No-Chinese-intrusion-

since-2010-only-transgressions-Govt /articleshow/40457901.cms, accessed on 27 June 2019

130. Ibid
131. Ibid
132. Ibid
133. Transgressions by Chinese troops reduced by over 50 per cent, while India builds up border infrastructure, *The Economic Times*, July 14, 2018, available at https://economictimes.indiatimes.com/news/defence/transgressions-by-chinese-troops-reduced-while-india-builds-up-border-infrastructure/articleshow/55030824.cms?from=mdr, accessed on 27 June 2019
134. Ibid
135. Doklam was tip of the iceberg! '426 Chinese incursions into India in 2017', *Business Standard*, web edition, Feb 6 2018, available at https://www.business-standard.com/article/current-affairs/doklam-was-tip-of-the-iceberg-426-chinese-incursions-into-india-in-2017-118020501172_1.html, accessed on 28 June 2018
136. Ibid
137. One year after Doklam, China intrudes 400m into Ladakh, *Times of India*, Web edition, 14 Aug. 2018, available at http://timesofindia.indiatimes.com/articleshow/65390945.cms?utm source=contentofinterest&utm_medium=text&utm_campaign=cppst, accessed on 26 June 2019
138. China Transgressed Into Indian Territory 30 Times in May This Year, *News 18*, web article, 24 July 2018, available at https://www.news18.com/news/india/china-transgressed-into-indian-territory-30-times-in-may-this-year-1821413.html, accessed on 28 June 2018
139. Chinese attempt to build a track on the Indian side of the Line of Actual Control (LAC) in Tuting area of Arunachal Pradesh, *Indian Express*, web edition, Jan 4, 2018, available at https://indianexpress.com/article/india/indian-forces-army-foil-chinese-attempt-to-build-track-in-arunachal-doklam-5010552/, accessed on 28 June 2019
    For more details see, Press Information Bureau of India release, Chinese Road in Arunachal Pradesh, Feb 5, 2018, available at http://www.pib.nic.in/Pressreleaseshare.aspx?PRID=1519149, accessed on 29 June 2019
140. Indian Army foils Chinese incursion in Ladakh, *Live Mint*, Aug 15, 2017, available at https://www.livemint.com/Politics/muaHDHjOIyHsvYJqIOre1J/Indian-Army-foils-Chinese-incursion-in-Ladakh.html accessed on June 30, 2019
141. Neville Maxwell, "Why the Sino-Indian Border Dispute is Still Unresolved after 50 Years: A Recapitulation," *China Report* 47:2(2011), SAGE Publications, 2011,, pp.75 and 81
142. M Taylor Fravel, "Regime Insecurity and International Cooperation - Explaining China's Compromises in Territorial Disputes," *International Security*, vol. 30,no. 2 (President and Fellows of Harvard College and the Massachusetts Institute of Technology, Fall 2005), pp. 46-83
143. Table 1, *China's Territorial Dispute: 1949-2005*, For more details about the territorial disputes of China, see M. Taylor Fravel, "Regime Insecurity and International Cooperation - Explaining China's Compromises in Territorial Disputes," *International Security*, vol. 30, no. 2 (President and Fellows of Harvard College and the Massachusetts Institute of Technology, Fall 2005), pp. 46-83.
144. "Arunachal Pradesh is our Territory: Chinese Envoy," November 14, 2006 at http://www.rediff.com/news/2006/nov/14china.htm, accessed on June 30, 2019
145. "China denies visa to IAS officer from Arunachal," *The Financial Express*, May 26, 2007 at http://www.financialexpress.com/news/China-denies-visa-to-IAS-officer-from-Arunachal/

200132/, accessed on June 30, 2019

146. China blocks ADB India loan plan, in *Financial Times,* web edition 11 April 2009, available at https://www.ft.com/content/033935c2-25e4-11de-be57-00144feabdc0, accessed on 29 June 2019
147. David Lai, Learning from the stones: A GO approach to Mastering China's Strategic Concept 'Shi', available at http://www.carlisle.army.mil/ssia, ccessed on 14 March 2109

5

Countering Collusive Hybrid Threat to India

"The greatest danger to our Nation continues to be external aggression with collusive threat from adversaries as well as internal instability due to state/ non-state sponsored proxy war, thus imposing multi front challenges in various dimensions, ranging from warfare in the non-contact domain, to an all-out full spectrum war."[1]

—Indian Army Land Warfare Doctrine-2018

ABSTRACT

This chapter will give an overview of security challenges to India with specific focus on the internal and emerging fault lines within country, which makes any state more vulnerable to hybrid threats. The contours of collusive hybrid threat have been analysed. This chapter argues that there is a need to build deterrence against any hybrid threat. The measures recommended to counter the collusive hybrid threat in the future have been covered in this chapter.

The Indian Army 'land warfare doctrine 2018' defines the primary security threat to the nation as a collusive threat from adversaries in conjunction with state-sponsored terrorism. The security challenges will spread across kinetic as well as non-kinetic domains. The threat of use of conventional force by both adversaries to resolve border issues, while simultaneously exploiting internal fault lines by sub-conventional, non-kinetic means to keep India engaged is based on well documented live precedents. Hostile neighbourhood and an increasingly complex security scenario is reality that cannot be ignored. What are the options for India? Is there a need to reform the current security arrangements in India? This chapter attempts to define the complex phenomena of Hybrid Warfare in the Indian context and gives an overview of security challenges to India with specific focus on the internal and emerging fault lines within the country, which makes the nation more vulnerable to hybrid threats. The contours of collusive hybrid threat have also been analysed in this chapter. The need to build deterrence against hybrid threats and measures recommended to *counter* the collusive hybrid threat in the future has been covered in this chapter. A case study of the 'Total Defence' concept being followed by Singapore has been analysed to seek inputs for drafting the recommendations.

Defining Hybrid Warfare in Indian Context

In chapters 1 and 2, the transformation of the character of war and the emerging prominence of hybrid war was covered. Chapters 3 and 4 chart the character of hybrid threats that India has been facing since independence from its traditional enemies. Based on the analysis of chapters 1 to 4, the recommended definition of hybrid warfare in the Indian context is given below:

> "*Hybrid Warfare can be defined as* ***armed*** *or* ***aggressive actions*** *by state and/or non-state actors* ***against a state*** *to achieve* ***stated political objectives,*** *actively supported by offensive stratagems in the diplomatic, economic, cyber, space, information and media domains undertaken simultaneously and persistently in the long term.*"
>
> *The successful use of* ***offensive stratagems*** *may in cases result in* ***achieving political objectives*** *which may preclude armed action or restrict it for a short time thus winning the war without fighting.*

The salient aspect of the concept of hybrid warfare is that it has to armed or aggressive action against a state, with stated political objectives. Only the use of non-kinetic means like diplomatic, information, cyber, psychological or media will qualify to be called hybrid warfare only if they are aimed against a state with a stated political objective. These non-kinetic means are being regularly used by transnational/multinational organisations against each other. In such scenarios, these stratagems will not fall under the domain of hybrid warfare.

Security Challenges to India

India is faced with a number of security challenges which can be divided into conventional and unconventional. Conventional security challenges like traditional animosity with Pakistan and increasing threats from China is well known and has also been discussed earlier in chapters 3 and 4, respectively. The line dividing conventional and unconventional security challenges is increasingly getting blurred. Hybrid security challenges involve a mix of conventional and non-conventional security challenges manifesting in combination. In this chapter, the non-conventional security challenges that India is facing along with the emerging security challenges that may translate into hybrid threat only have been covered:

Vulnerable Land Borders. India has long land borders which traverse through the most varied forms of terrain. The land borders run from the marshes of the Rann of Kutch in Gujarat, sandy deserts in Rajasthan, riverine terrain of Punjab, mountains and glaciated terrain in Jammu and Kashmir and Uttar Pradesh, to mountainous, tropical jungles and the Sundarbans delta in the North East. The land borders with China and Pakistan are not settled; those with Myanmar and Bangladesh are partially settled. The unsettled borders, with inhospitable terrain and a not so stable neighbourhood, make border management a major security challenge. India shares borders with seven countries, the details of which are is given in Table 5.1. The challenges of managing the land borders with China, Bangladesh, Myanmar and Pakistan are discussed in brief[2] as under:

- **Land Border with China.** India shares approximately 3,488 km of land border with China, which traverses through mountain ranges of more than 18,000 feet, glaciated terrain, snow deserts and thick jungles. The terrain is extremely rugged and inhospitable. The border between the two nations is not defined or delimited. However, both have agreed on the Line of Actual Control. Each has its own understanding of the LAC and carry out patrolling up to it. This leads to confrontation between the troops of both the countries, which has been discussed in detail in chapter 4.
- **Land Border with Pakistan.** The border with Pakistan consists of the international border in Gujarat, Rajasthan, Punjab and Jammu & Kashmir and the Line of Control (LoC) and Actual Ground Position Line (AGPL) in Jammu & Kashmir. The terrain along the LoC is extremely rugged, covered with dense forests and is prone to infiltration. The border is completely fenced and guarded. However, the rugged terrain and inclement weather and the support of the Pakistan army in terms of launch pads facilitates infiltration.
- **Land Border with Bangladesh.** The land border with Bangladesh is porous and one of the most complex. In West Bengal, the Sundarbans

mangroves, rivers and small islands (called chars) with infested riverine terrain makes border guarding very difficult. The terrain in Assam, Meghalaya, Tripura and Mizoram is mountainous, rugged and thickly forested. Illegal migration through the porous borders is one of the major security challenges. The porous border and high density of population along the Assam and northern part of the border with West Bengal, makes it an easy route of infiltration/exfiltration for insurgent groups like ULFA, NSCN-IM and several others operating in the north eastern part of India.[3] There have been numerous reports of the ISI of Pakistan also exploiting the porous India–Bangladesh border for its nefarious designs against India.

- **Land Border with Myanmar.** The land border between Myanmar and India is approximately 1,640 km and traverses through mountainous and thick jungle terrain. Again the terrain is extremely rugged and facilitates undetected movement of personnel. The border has been a favoured route for infiltration and exfiltration of insurgent groups active in the North East.

Table 5.1: India's Land Borders

Sl.No.	*Border with*	*Length (Km)*	*Type of Terrain*	*States*
1	China	3488	Mountain, Glaciated & Jungle	J&K, Himachal, Uttarakhand, Sikkim & Arunachal Pradesh
2	Pakistan	3323	Marshy land, Desert, Riverine, Hills, Mountain & Glaciated	Gujarat, Rajasthan, Punjab & J&K
3	Bangladesh	4096	Delta & Jungle	West Bengal, Meghalaya, Assam, Tripura & Mizoram
4	Nepal	1751	Mountain &Tarai	Uttarakhand, Bihar, West Bengal, Sikkim & Uttar Pradesh
5	Bhutan	699	Mountain & Jungle	West Bengal, Sikkim, Assam & Arunachal Pradesh
6	Myanmar	1640	Mountain & Jungle	Manipur, Mizoram, Nagaland & Arunachal Pradesh
7	Afghanistan	106	Mountainous	Jammu and Kashmir

Source: Border Management Division of Ministry of Home Affairs

Coastal Security. India has a long and complex maritime boundary measuring 7,516.6 km. The maritime boundary is contiguous to the mainland and the islands in the Bay of Bengal in the East, the Indian Ocean on the South and the Arabian Sea on the West.[4] There are a total of 1,382 offshore islands.[5] India has been able to resolve all the maritime boundary issues with its neighbours except with Pakistan on Sir Creek. The last maritime boundary issue resolved was with Bangladesh in 2014. Coastal security has an important role in the overall security matrix of any nation. Prevention of smuggling, narcotics networks, illegal arms sale, illegal fishing and movement of refugees are some aspects which make coastal security a complex issue. The Mumbai terrorist attacks of 2008 showed up the hollowness in our coastal security mechanism. The terrorists who carried out the attacks had taken the sea route from the coast of Karachi to reach Mumbai undetected. Even, in the 1993 Mumbai serial blasts, the explosive used is reported to have been smuggled via the Raighad coastal belt.[6] There have been reports of the LeT likely to have used Lakshadweep for infiltration of terrorists and weapons for Jammu and Kashmir.[7] The involvement of a large number of government departments like Customs, fisheries, ports & shipping, environment, state governments and home makes formulation, implementation and coordination a complex issue.

Table 5.2: India's Coast Line

Sl.No.	*State*	*Length of Coast Line (in km)*
1	Gujarat	1,214.70
2	Maharashtra	652.60
3	Goa	101.00
4	Karnataka	208.00
5	Kerala	569.70
6	Tamil Nadu	906.90
7	Andhra Pradesh	973.70
8	Odisha	476.70
9	West Bengal	157.50
10	Daman & Diu	42.50
11	Lakshadweep	132.00
12	Puducherry	47.60
13	Andaman & Nicobar Islands	1,962.00

Source: Annual Report of 2018-19, Ministry of Home Affairs

Turbulent Neighbourhood. China and Pakistan may be termed traditional threats to India. However, the troubled and unstable neighbourhood is potentially a bigger challenge when the hybrid threat is the context. As the Indian sub-continent is tied together with shared history and culture, political stability amongst the neighbouring countries is crucial to in the security calculus of India. Unfortunately, almost all of these countries except for Bhutan have gone through an extremely turbulent political and instable economic past. Moreover, the asymmetry in size, economy, GDP, infrastructure, social development, technological threshold and armed forces tends to create a feeling of insecurity when compared with India. In order to balance or overcome this insecurity, these nations have been looking outwards for support and this opportunity has been utilised by China.[8] The current state of the immediate neighbourhood and how the prevailing situation adversely affects India's security environment is discussed as under:

- **Afghanistan.** Historically, Afghanistan has been the battle arena for major and regional powers. In recent times, consequent to the Soviet invasion in the 1980s and creation of the Mujahedeen to fight the Russian forces, has resulted in Afghanistan being the epicentre of Islamic the Jihadi movement.[9] The GWOT for the USA in Afghanistan is eventually turning out to be the longest war. The irony is that even after fighting Islamic terror for 18 years, the USA is today talking with the leader of the Islamic terror movement in Afghanistan. According to a former envoy to Afghanistan, Shri Vivek Katju, *"The Taliban have occupied the space in the peace dialogue, which should be legitimately that of the state or the government,"*[10] While the US-led peace process has had seven rounds of talks with the Taliban, there is no official participation from the Government of Afghanistan, hence it questions the legitimacy and acceptance by the people of Afghanistan. A stable and peaceful Afghanistan is crucial for the long-term security of India and the region. The possibility of a civil war in Afghanistan after the exit of the USA cannot be ruled out, as the Afghan National Defence Security Forces (ANDSF) which has a considerable non-Pushto representation is unlikely to accept the rule of the Taliban.[11] Such a scenario would be of most concern for India.
- **Bangladesh.** India surrounds Bangladesh from three sides and shares a common border of 4,000 km. Therefore, any social, political, religious or economic development affects India directly. Illegal migration from Bangladesh is the biggest security challenge for India. The increasing population and lack of resources in Bangladesh has been the instrumental reason for illegal migration into India. The growth of the population is explained by Figure 5.1. The Home Sinister for State, in 2016 informed the Rajya Sabha in reply to a written question, that around 20 million Bangladeshi migrants were living in India. The illegal migration has also severely affected the demographic balance in the bordering states of Assam, West Bengal and Tripura.[12] The migrant Muslim population is always

vulnerable to influence by radical, fundamentalist and terrorist groups. The increasing Chinese influence in Bangladesh is also a security concern for India.

Figure 5.1

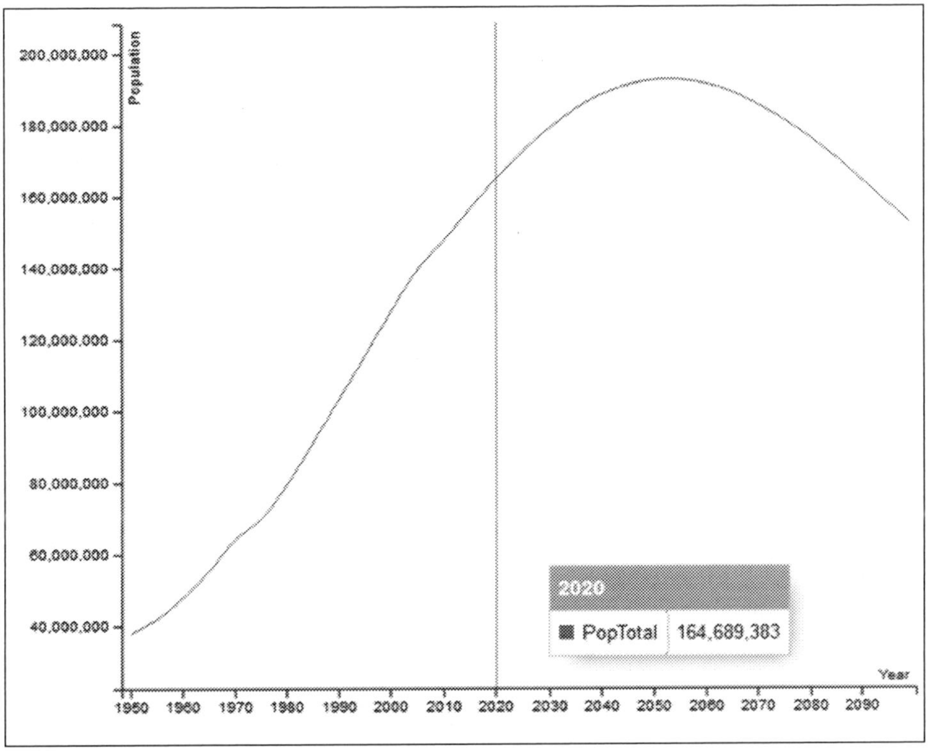

Source: Population Growth Chart of Bangladesh[13]

- **Nepal**
  - Nepal and India are tied geographically and share close cultural bonds. The Treaty of 1950 provides Nepal special status, which legislates an open border between the two countries and allows Nepalese citizens to work and live in India. Nepal while being dependent on India for most of its supplies, has also been playing a balancing act between India and China. To India's discomfort, the last few years has seen unprecedented growth in Nepal-China relations. The relations between India and Nepal went on a downward spiral from 2015-16 onwards after the five-month long economic blockade. In 2018, during Prime Minister Oli's visit to Beijing, deals worth $2.4 billion

were signed, with the focus on infrastructure and energy projects.[14] Recently, Nepal has shown interest in joining China's BRI project and is already working on the Keyung-Kathmandu railway line, which will be connected to the Chinese Qinghai-Tibet railway at Xigatse (Xigase). Reportedly, the feasibility study for the project has been completed and work on the DPR is about to commence.[15] In addition, Nepal's President, Bidya D. Bhandari, during his recent visit to China, signed a protocol to use China's sea and road infrastructure for trade with third countries, which was a fallout of the Transit and Transport Treaty of 2016. China is also involved in the construction of an international airport at Pokhra. The project is being financed by the EXIM Bank of China and is being constructed by China CAMC Engineering. One of the manifestations of the increasing closeness with China was visible in 2018, when Nepal withdrew from the first joint military training exercise of BIMSTEC nations in September 2018.[16]

- Nepal has also emerged as the hub of the ISI's nefarious activities. The increasing vigilance on the international border, LoC and construction of fences has made infiltration difficult for terrorist groups operating in Pakistan. Hence, the ISI has started exploiting the open borders between Nepal and India for inducting their cadres and weapons/ammunitions into Kashmir.[17] The India-Nepal 'open' border runs across 20 districts in five Indian states: Bihar, Uttar Pradesh, Sikkim, West Bengal and Uttaranchal. The ISI has been using Nepal as its regional command and control centre for activities in India and the India-Nepal border of Uttar Pradesh and Bihar for movement of their sources.[18] The role of the Pakistan embassy in Nepal also came to light in 2000 when Nepalese police apprehended RDX sealed in diplomatic bags of the embassy.[19] In April 2018, the Indian Consulate office wall was damaged by a bomb which was reportedly planted by a communist group financed by the ISI.[20] In 2018, the Intelligence Bureau had provided the Nepal government a list five ISI operatives in Biratnagar and Kathmandu, who had been actively involved in various illegal activities in India. Another emerging challenge to India is the changing demography along the Terai region of Nepal. As per media reports, in the last 15 years more than 100,000 persons have embraced Islam in the Terai region.

According to a report on India's border management by the MHA in 2000, there are 343 mosques, 300 madrassas and 17 mosques-cum-madrassas within 10 kilometres of the border on the Indian side. On the Nepal side, there are 282 mosques, 181 madrassas and eight mosques-cum-madrassas. The report highlights that these madrassas and mosques receive funds from Gulf countries.

- **Myanmar.** Myanmar is one of the most important players in India's 'Look East Policy', being the bridge between South Asia and South East Asia. It not only connects India physically but also culturally, historically, spiritually and economically. India's relationship with Myanmar has been more stable when compared with other neighbours. However, there are areas of concern in the context of security. The porous, mountainous and thickly forested border is used by insurgent groups active in the North East. Myanmar also remains a golden route for the smuggling of arms, drugs and contraband. Myanmar has its own insurgency problems on the eastern border with China, where the insurgency is reported to be supported by China,[21] especially the United Wa State Army (USWA) which is the largest insurgent group fighting against the Myanmar government.[22] The recent crackdown on the Rohingya community has created a new security problem for India. This Muslim ethnic group of refugees is vulnerable to be influenced by radical terrorism and may end up being targets for recruitment by Islamic terror groups. Chinese mega infrastructure projects in Myanmar are another security concern for India. A debt-trapped Myanmar under Chinese loans is not a positive happy scenario for India.
- **Sri Lanka.** The historical, mythological and cultural ties between India and Sri Lanka are well known. While, the social fabric of Sri Lanka has been traditionally affected by the Sinhala-Tamil conflict, in the last few years Islamic and Christian identity has further complicated the internal dynamics. The ethnic mosaic of Sri Lanka is dominated by Buddhists 70 percent, followed by Tamils 12 percent, Muslims 10 percent and Catholics 6 percent.[23] The recent targeting of three Catholic churches indicate that the Christian minority and the international audience were the targets. The ISIS, claiming responsibility for the attacks, shows the vulnerability of Sri Lanka in the spread of radical terrorism. Sri Lanka is critical to India's security parameter due to its strategic position in the Indian Ocean. Chinese inroads into Sri Lanka and the annexation of the Hambantota

port through economic coercion have been discussed in Chapter 4. The increasing China-Sri Lanka collaboration in the military field is also a matter of concern for India.

Existing Internal Fault Lines

The COAS, General Bipin Rawat, while delivering the keynote address at a national seminar on Internal Security organised by CLAWS in November 2018, had said that "*internal security situation remains the biggest challenge for the nation, and the reason why the internal conflicts persists even after so many decades is due to their external linkages.*" This brings out two important issues; first, 'internal security is the biggest challenge' and second, 'external linkages'. The most critical vulnerability in hybrid warfare is the presence of an internal fault line, which can be exploited by the adversary. India comprises innumerable faiths and sub-faiths, cultures, castes, languages, regions and customs. According to an estimate, 1,660 local languages are spoken in India of which 22 are recognized as official. There are nearly 3,000 castes and sub-castes, and nearly all faiths and known religions are practised. This makes diversity the biggest strength.[24] However, this strength at times also gets exploited as the diversity creates fault lines and vulnerabilities. Such fault lines, which are affecting internal stability, are discussed below.

- **Secessionist Movement of Jammu and Kashmir.** The princely state of Jammu and Kashmir joined India through the Instrument of Accession. However, Pakistan considers it as an unfinished agenda.[25] While Pakistan may not have been successful in the three wars that it initiated,, the proxy war that it instigated based on the religious fault line has kept the level of violence and terrorism high in the Kashmir Valley. The proxy war initiated by Pakistan in Jammu and Kashmir has been covered in detail in Chapter 3. The problem is that Jammu and Kashmir is the biggest security challenge for India, as it has both external and internal security angles.
- **Insurgency in North East.** North Eastern India comprises eight states (including Sikkim) and accounts for approximately 4 percent of the population of India, 8 percent of the total land, 40 percent of water resources but contributes 0.1 percent of India's GDP. These statistics describe the core of the issue—years of neglect by successive governments at the Centre since independence, geo-spatial segregation which was further increased due to the partition and creation of East Pakistan/ Bangladesh, diversity in tribes (more than 200 tribes in the region) and culture and language are some of the reasons which led to insurgency

movements. The Naga insurgency was the first., Later, almost all states except for Arunachal and Sikkim had some insurgency movement. However, in the last few years the level of violence has been under control and most of the insurgencies have been successfully neutralized. The situation still remains vulnerable to external interference. As discussed in Chapter 4, China will continue to cultivate these groups by providing covert support so that it retains leverage to exploit them at the time of its choosing.

- **Left Wing Extremism.** Left Wing Extremism (LWE) was described by former Prime Minister Dr. Manmohan Singh as the biggest threat to the internal security of India.[26] The LWE movement can be traced back to 1969 to the village of Naxalbari in West Bengal when Charu Mazumdar founded the Communist Party of India (M-L), with the aim to overthrow the country's democratic system. The movement soon spread to other states like Andhra Pradesh, Madhya Pradesh, Chhattisgarh, Jharkhand, Bihar and Orissa. The violence profile in the LWE-affected areas is give in Figure 5.2

Figure 5.2: Casualties in LWE affected areas

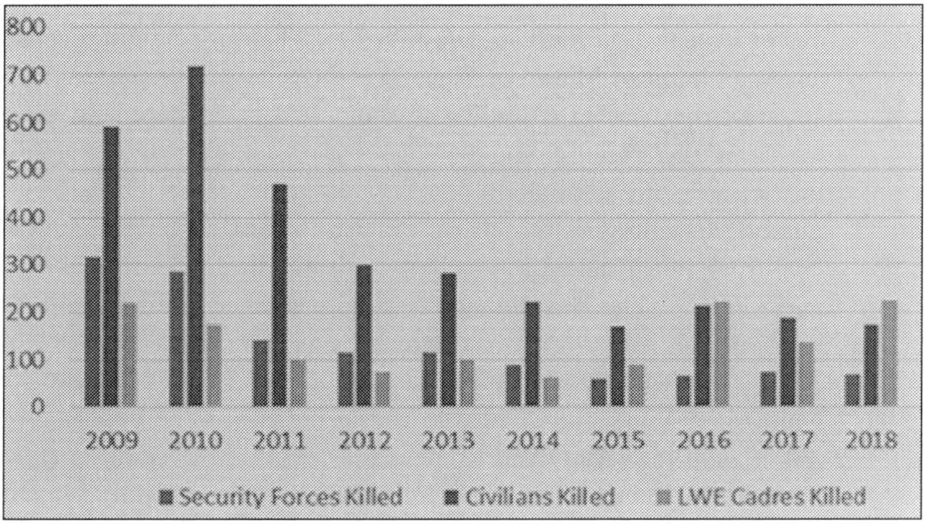

Source: Parliamentary Question[27]

The situation in LWE areas has been improving since 2014 when the casualties of civilians and security forces have been low while for LWE cadres it has been high. The LWE problem essentially is an issue of governance; the solution lies in development of the area, providing employment to affected people and economic engagement. The role of security forces is limited to provide a secure environment for government agencies to function.

- **Communalism.** India is a secular country according to the Constitution of India. Article 15 of the Constitution, grants all citizens the right to equality and there is no place for discrimination on the basis of religion, caste, language, ethnicity or gender. Everyone has a right to practise a religion of his/her choice. However, in practice, there is communalism existing on the ground, and this has led to a number of serious communal riots since independence. The fault line based on religion is getting deeper day by day. This has also manifested in the vote bank politics of various region and castes. Today, caste/sub-caste dynamics are deciding the outcome of elections rather than the merit of the candidate. The communal fault line has been used successfully by our adversaries in the past.
- **Structural and Systemic Fault Lines of Indian Governance System.** The lack of coordination between the civil and military establishments and between the services itself has been highlighted in a number of reports and by scholars. The most notable is the Kargil Review committee report and the Group of Ministers (GoM) Report. The lack of coordination amongst the services is leading to individual prioritisation in force modernisation, planning and training thereby leading to duplication and inefficient resource utilisation. India's weakness on joint action calls into question its ability to formulate a coherent tri-services response to the 'two-front threat' posed by ongoing territorial disputes with Pakistan and China.[28] The coordination and convergence amongst various ministries on issues of national interest is of grave concern. A case in point is that while the MEA and the MoD consider China a threat, the Ministry of Trade and Commerce considers China as an opportunity. While the MEA, MoD and MHA are reviewing the decision to allow Chinese communication companies like Huawei for bidding in the tendering process for 5G in India, the majority of computer and networking hardware in these ministries is already Chinese. Such fundamental differences and lack of convergence between the ministries adversely hamper the overall security framework of the nation.

> *The complete lack of sharing of intelligence between various organisations was a big problem. The army, navy and other agencies were not sharing their intelligence inputs with each other.*
>
> **—P.C. Chidambaram, post 26/11**

- **Integration of the Intelligence Apparatus.** The first holistic review of National Security was probably after the Kargil War. Towards implementation of the report of the Kargil Review Committee, Government had set up four task forces, one of which was on Intelligence headed by Shri G.C. Saxena.[29]

The task force report was never declassified due to obvious reasons of national security. However, two new intelligence set-ups were created, i.e., National Technical Research Organisation (NTRO) and the Defence Intelligence Agency (DIA). Prior to this, there were the Intelligence Bureau and the Research and Analysis Wing (R&AW). In addition, there are other Central and State Intelligence agencies like the Intelligence agencies of the three arms of the Armed Forces. With the kind of complex, heterogeneous and dynamic security environment India has, timely intelligence becomes the key to pre-empt proxy war, radicalization, cross border terrorism, etc. The integration of the Intelligence agencies or the synergy between them has affected the overall response of the nation against non-conventional threats. Post Kargil, the Multi-Agency Centre (MAC) was set up as per the recommendations of the GoM.[30] These MACs have been further connected with the Subsidiary Multi-Agency Centre (SMAC). The MAC-SMAC Network is further linked to other law enforcing agencies like the State police forces. By December 2018, 429 SMAC nodes and 251 district police offices were connected to the MAC-SMAC network.[31] However, the shortage of trained personnel and resistance to change has been affecting the overall output of the system. In addition, the tendency of guarding its own turf by each agency is another major hurdle in the sharing of intelligence. The entire system today is dependent on interpersonal relationship between the officers of the various Intelligence agencies rather than the protocol of flow of information.[32] The 26/11 Mumbai attacks led to the creation of the National Investigation Agency (NIA), National Counter Terrorism Centre (NCTC) and a national computerised information-sharing network known as NATGRID. However, most of these agencies did not evolve as conceptualised. Both the NIA and the NCTC were stuck in the quagmire of Centre-State political dynamics and finally NCTC, which was to be the pivot of counter-terrorist operations, was placed under the Intelligence Bureau. According to a recent press release of the MHA in September 2109, the NATGRID project is to be ready by early 2020.[33] According to the release, "*The NATGRID will have data related to all immigration entry and exit, banking and financial transactions, credit card purchases, telecom, individual taxpayers, air flyers, train travellers besides others to generate intelligence inputs.*"[34] In phase I, 10 central agencies will be able to access the data. These is a positive development. However, more efforts are required to synergise the effort of the intelligence agencies, as these are the first line of the defence against hybrid war.

Emerging Fault Lines/Vulnerabilities

The changing social, political and economic dynamics is creating new fault lines and thereby new security challenges. New technologies like social media and digital financial instruments are impacting societal behaviour. These may end up creating new fault lines and vulnerabilities. Such emerging trends have been discussed in this chapter.

- **Increasing Economic Disparity.** The benefits of economic growth in India have not been shared evenly. Rather the disparity between the well-to-do and not so well-to-do has widened. A report by Oxfam, quoted by *Business Today*, brings out that the top 1 percent in India bagged 73 percent of the total wealth.[35] In fact, India has emerged as the country with the highest increase in the top 1 percent income share concentration over the past 30 years. The report indicates that while the Indian GDP is growing, the poor are getting poorer. The increasing disparity creates a desperation and frustration amongst the poor class. This creates a vulnerability which can easily be exploited in hybrid warfare.
- **Regional Tendencies.** The issues of strategic importance are being influenced by regional dynamics. Regional political sensitive issues like the Teesta water sharing in West Bengal or issues with respect to the Tamils in Sri Lanka are affecting relationships or management of strategic concerns with Bangladesh and Sri Lanka. Prominent political leader of Tamil Nadu, Mr. Stalin, in 2018 had raised the issue of 'Dravida Nadu' as a separate nation consisting of South Indian states.[36] A case in point is the issue of the 'Kartarpur Sahab Corridor'. How sensitive is the proposal in the context of internal security and for the revival of the Khalistan movement is a matter of discussion. However, as the proposal has been made by Pakistan, now the deliberations and consideration of the proposal

will be influenced by factors of emotional connect of the people of Punjab, the current government and the government at the centre. The outcome of the next elections will override the strategic importance of the issue. Such regional tendencies are detrimental to national security interests.

- **Illegal Migration and Changing Demography in Border States.** Any bordering state has to tackle the problem of illegal migration. However, the situation is acute in states bordering Bangladesh. The magnitude of the problem of illegal migration in the states of Assam and West Bengal can be understood by the fact that in both states the core issue for elections was illegal migration. In both the states there is no census for the illegal migrants from Bangladesh. However, as most of the migrants are likely to be Muslims, the increase in their percentage can be taken as an indicator to assess the enormity of the issue. In West Bengal, at the time of Partition, the Muslim population stood at 12 percent; by 2011, West Bengal's Muslim population has increased to 27 percent (up to 63 percent in some districts). According to the 2011 Census data, Assam recorded the highest increase in the share of Muslim from 30.9 percent (2001) of the state's population to 34.2 percent (2011). Any democracy is vulnerable to vote bank politics. The migration into West Bengal has changed the way the elections are being fought in the state. The migrants being the most lucrative vote bank, manifestoes of political parties today are influenced by the share of vote that Bangladeshis are controlling, who technically are illegal. Apart from affecting the political dynamics, it creates serious security implications. The refugees are always vulnerable to fall into the trap of terrorist and anti-national organizations. An *India Today* report quotes an investigation report of the NIA which brought out that '*there are 58 terror modules operating in West Bengal with 180 Bangladeshi nationals and are likely to be members of Jamaat-ul-Mujahideen Bangladesh (JMB). These terror modules are likely to be linked with terrorist groups from across the border having connections with al-Qaeda, LeT and HUJI*'.[37] The security-related issues with respect to illegal migrants or refugees is not restricted to the border states alone;, in the urban centres they are generally residing in pockets generally in slums under unhygienic conditions which are nurseries for recruitment by terrorist organisations.
- **Cyber Security.** In the increasing dependence on networks and information, technology has created a new fault line, and which is most

sensitive. Lack of cyber security hygiene, training and consciousness puts the entire network at risk and makes it vulnerable to cyber warfare. The lack of a cyber-security doctrine or a framework or clear guidelines makes decision making flawed or focused at the short term. A case in point is the launching of 5G in India. While the strategic community, the MEA, MoD or MHA may consider Huawei of China as a security concern like many other countries have due to Huawei's closeness to the PLA and Communist Party of China, at the same time, the backbone communication infrastructure of MTNL or BSN is using Chinese computer hardware from companies like ZTE and Huawei. Hence, hypothetically restricting Huawei only for 5G may not make much sense. As per an Indian Government study in August 2018, 5G has some very serious national security implications.[38] But the Steering Committee on 'Making India 5G Ready' does not have any representative from the defence or intelligence community.[39] Cyber warfare is one of the most preferred instruments of hybrid war. In June 2019, during the Iran-US standoff and consequent to Iran not abiding by the nuclear deal, the USA preferred to use cyber-attacks in reply to Iran shooting down a US drone.[40] The cyber-attack disabled the Iranian computer systems that controlled its rocket and missile launchers. In Chapter 4, China's focused attempt to build a formidable cyber offensive and defensive capability has been discussed. In India, setting up of the National Cyber Coordination Centre (NCCC) is a big step towards creating a cyber-defensive mechanism by building a capability to scan the country's web traffic to detect cyber security threats. The NCCC is primarily a coordination centre; it has more of an advisory role and its power to enforce discipline on various organisations is still not very clear. The cyber domain is a critical vulnerability which spans across all domains.

- **Digital Financial Instruments.** The advent of new digital financial instruments like Bitcoin, Litecoin (LTC), Ethereum (ETH), Zcash (ZEC) and other crypto currencies is making it difficult to track financial transactions. These digital financial instruments have certain unique characteristics which facilitate transnational financial transactions. These characteristics are anonymity, usability, security, acceptance, reliability and volume.[41] These characteristics are explained in Table 5.3. Characteristics like anonymity of both sender and receiver, security of transaction, acceptance across maximum countries and the large volume of single transactions makes it an attractive option for transnational terrorist organisations, while making it a challenge for security agencies to counter terror financing. As Bitcoin is not recognized in India, hence it becomes an even more attractive instrument for terror organisations in India.[42] There is a need to devise mechanisms to address this issue.

Table 5.3: Characteristics of Crypto Currencies

Sl.No.	*Character*	*Property*
1	Anonymity	The ability to hide and protect the identity of the user
2	Usability	The ease with which the user can conduct transactions and manage his or her own currency
3	Security	The degree to which the crypto currency infrastructure secures the confidentiality, integrity, and accuracy of transactions and user accounts
4	Acceptance	The degree to which the currency is accepted by a user community as well as the size of the community of users
5	Reliability	The speed and availability of transactions, as viewed by users
6	Volume	The time-averaged aggregate size of transactions in the crypto currency infrastructure

Source: Compiled from the Rand Report

"*Collusive threat from China and Pakistan to India implies both countries acting in secret to achieve a fraudulent, illegal, or deceitful goal and being engaged in secret or hidden avowed goals vis-à-vis India*"[43]

—General (Retd) V. P. Malik, PVSM, AVSM
Former COAS

Collusive Hybrid Threat to India

The ongoing hybrid war or future threat from Pakistan and China have been discussed in detail in chapters 3 and 4, respectively. This part will analyse the collusion in the hybrid domain between China and Pakistan. Since independence, during the wars that India has had with Pakistan or China, there is no major incident of direct collusive support amongst the two nations against India in any of them. There was one such incident during the Kargil conflict Pakistan when the PLA tried to transgress in Yangtse sub sector which led to a faceoff with the Indian Army, which lasted almost close to 30 days.[44] In retrospect, this particular incident would have prevented moving of forces from the Eastern sector, if India wanted to. It indicates collusion between China and Pakistan. This is a hypothetical analysis, as India did not cross the LoC during the Kargil conflict; it captured the posts which were illegally occupied by Pakistani troops. There are many other incidents where the support of China to Pakistan would fall in the realm of hybrid war. The important ones are given below:

> *"If you subtract China's help from Pakistan's nuclear programme, there is no Pakistani nuclear weapons programme."*[45]
>
> **—Gary Milhollin, US Nuclear Analyst**

- **Assistance in Pakistan's Nuclear Program.** The Chinese assistance to Pakistan for developing its nuclear bomb commenced with the assurance given by Mao to Zulfiqar Ali Bhutto in 1976, when the later visited China.[46] It was a two way process. While China also benefitted with designs of nuclear enrichment plants that were stolen by Pakistani nuclear scientist A.Q. Khan, it provided Pakistan with designs of nuclear bombs.[47] China supplied the critical components for the nuclear weapon program like Ring Magnets required for making weapons-grade enriched uranium, Tritium to boost the yield of atomic weapons, heavy water needed to operate the plutonium production reactor and many more.[48] China also provided the Khushab reactor, from which Pakistan is known to have drawn plutonium for weapon production.[49]. It was not only material and technological support, Chinese also trained Pakistani nuclear scientists who were taken to China for formal training.[50]
- **Assistance in Pakistan's Missile Development.** China has assisted Pakistan in developing nuclear weapons, but larger assistance has been in the field of the missile development program to deliver nuclear warheads. According to CIA reports, China has been assisting Pakistan's missile program since the early 90s. China supplied 34 complete M-11 missiles and subsequently helped setting up of a factory for the production of M-11 missiles at Tarwanah, near Rawalpindi.[51] Later, China assisted in the serial production of SHAHEEN-I (SRBM) and SHAHEEN-II (MRBM) also. China while claiming to be aligned to the MTCR regime is however not a signatory. Accordingly, China has been supplying material, and financial assistance to support Pakistan's the missile program.[52]
- **Chinese Support to Pakistan's Proxy War in Jammu and Kashmir.** The Chinese approach towards the Kashmir issue started tilting towards Pakistan from the early 60s. In 1963, Pakistan handed over 5,180 sq km of Indian Territory in Kashmir (Shaksgam Valley to China),[53] which was

under illegal occupation by Pakistan. Since then, China has been supporting Pakistan on the issue of Kashmir as a trusted all-weather friend. While there is no formal alliance, China's support to Pakistan in diplomatic forums can be quantified as that of more than an ally.[54] China has never highlighted Pakistan's role in terror; rather, it has always scuttled Indian efforts to raise this issue in various international forums. China's has even gone out of the way to prevent UN resolutions on banning known terrorists like Zaki-ur-Rehman Lakhvi, Hafiz Saeed and Masood Azhar. In order to keep Islamic terrorism under control in Xinxiang and to protect the investments on CPEC in Baluchistan, China may even get into collaboration or tie up with Islamic terrorist organisations like the LeT and JeM, and in bargain may provide them diplomatic support on the Kashmir issue. This particular argument needs to be seen in conjunction with reports in the Chinese media post 2016 on the Kashmir issue, wherein the Chinese print media played the narrative of '*excessive use of force by the Indian government to suppress local calls for autonomy*'.[55] Even on the terror strike on the CRPF convoy at Pulwama in J&K, some sections of Chinese print media opined that '*India should not blame Pakistan and China.*'[56] China will continue to play the double game in the region with reference to Islamic terrorism. Chinese hypocrisy in its stance is so blatant that on one hand it is brutally suppressing the Xinxiang freedom movement quoting the rise of Islamic extremism and on the other hand it provides all possible support to Pakistan for its proxy war against India.

- **Changing Dynamics in PoK.** China has had an interest in the Gilgit-Baltistan (GB) region for long; the Karakoram Highway (KKH) was built in 1982 through GB. The China Pakistan Economic Corridor (CPEC) has come up along the same alignment by upgrading the infrastructure along the erstwhile KKH. The manifestation of the CPEC has taken the Chinese involvement in PoK to the next level. With Chinese stakes having gone up due to the construction of Gwadar port, economic corridor, SEZ and deployment of thousands of Chinese workers, China cannot disengage itself from the dynamics of PoK. China has even deployed PLA troops in Gilgit and Pakistan has created a new formation for security of the CPEC. The infrastructure being built in PoK to support Chinese interests and investment will automatically boost the Pakistan Army's logistic capability and response mechanism. There has been a pro-independence movement in GB, due to misadministration and neglect of the region by the Pakistan Government over the last 70 years. Beijing, appreciating that any trouble in GB would threaten the CPEC, has been pressurising the Pakistan government to settle the GB issue. Hence, the Pakistani government is now likely to bring an ordinance to declare GB as the fifth province of Pakistan.[57] This indicates the control of Beijing over the internal affairs of Pakistan. The Western Highway in Tibet, CPEC and BRI projects from Xinxiang to Khunjerab Pass encircle the state of Jammu and Kashmir. The geostrategic importance of Kashmir is now being defined by the Chinese stake in the success of the BRI which is the show window of President Xi, and hence China–Pakistan collusion will only increase in the days to come.

Collusive Threat in Future. After the Balakot air strikes, there has been an improvement in the security situation in J&K. There has been a decrease in infiltration there by 43 percent in the first half of 2019.[58] However, this is cannot be taken as an indicator of any changing trend. Pakistan is unlikely to stop abetting terrorism and fighting the proxy war against India. While China may not overtly support Pakistan in its proxy war against India in the present context, it would continue assisting Pakistan diplomatically, militarily and economically in all its misadventures. China is unlikely to make any positive move to resolve the boundary issue; rather, it will keep the LAC active by orchestrating face-offs on it which will keep the Indian Army engaged. China will also retain leverage to influence the internal situation in North East India. The collusive hybrid threat in future is a reality which needs to be acknowledged now. India, while continuing to build its conventional capability simultaneously needs to guard the fault lines against being exploited by adversaries. There is also a need to counter hybrid warfare. In the next part of this chapter, the various options to counter hybrid warfare will be analysed.

Countering Hybrid Threat: Building Deterrence

Two nuclear-powered hostile neighbours is a geographical reality for India. Pakistan has attempted four wars, suffered decisive defeats in all of them and has now been engaged in a proxy war against India for more than 30 years. The Indian strategy of having a conventional credible deterrence has not deterred Pakistan to undertake such misadventures. Pakistan's ongoing proxy war is a clear indicator that India's strategy of deterrence by denial has not been successful.[59] The theory of deterrence is simple to understand in a conventional threat scenario and where the actors behave rationally.[60] This raises two important issues, which can be related to the behaviour of Pakistan. First being the rationality of fighting four wars and sustaining a 30–year-long proxy war, and second is about the nature of threat from Pakistan, which has never been purely conventional; rather, has always been in the domain of hybrid war.[61] Will the decisions of repeatedly undertaking war with India fall under irrationality? Who were the leaders when the decisions of going to war with India were taken? That takes us to even deeper questions like were the political leaders during the time of decision-making irrational? Or there were some other political pressures and internal factors which forced them to take such decisions? The analysis in Chapter 3, puts some of these questions in context. The political leadership has rarely been in the driving seat of India-Pakistan relations. It is the Pakistan army which has been the pivot of all decisions to undertake war against India since 1947-48 till 1999 (Kargil). So, while to a rational military analyst the decisions of repeatedly going to war with India may appear to be irrational, actually it is not so. As discussed in detail in Chapter 3, it is the strategic culture of the Pakistani Army, which defines the rationality or the necessity of the decisions of undertaking four wars and now fighting a proxy war against India.

The next important aspect in analysing the theory of deterrence is the 'threat', the defining or understanding of the 'threat' is the basis of the theory of deterrence. The nature of 'threat' defines the nature of deterrence to be adopted by the country. If the threat has an asymmetric component, then the deterrence becomes complex.[62] In asymmetric or sub-conventional warfare (insurgencies), the stronger power, while trying to create the deterrence against sub-conventional threat, is likely to fall in the 'deterrence trap.'[63] The weaker party/insurgents use violence against the civilian population and other vulnerable targets, thereby forcing the stronger power to react or escalate the situation or undertake measures which become counterproductive. It is the characteristic of sub-conventional war that it creates a vicious cycle for the stronger power, wherein the pre-emptive/ precautionary/prohibitory actions to deter the insurgents, creates more anguish for the local population, which is exploited by the insurgents, which follows further stringent measures by the stronger power to preserve the decision it has taken so far, which leads to further resentment and frustration in the population, and the situation eventually favours the insurgents rather than the stronger power. This is how the stronger power gets into a deterrence trap. The deterrence in sub-conventional warfare may end up being the problem itself rather than the solution.[64] However, the situation in Jammu and Kashmir is not purely home-grown insurgency. It is a proxy war by Pakistan, wherein non-state actors like the LeT, JeM, HM and other such terrorist groups are funded and supported by Pakistan. Hence, the focus of India should be more towards creating a deterrence which will prevent/discourage Pakistan to support the terrorist groups operating in Kashmir, rather than enforcing very strict deterrence measures against its own population. Hence, two issues emerge from the argument that Pakistan's behaviour is not irrational; rather it is an outcome of the deep-rooted beliefs of its strategic culture. Secondly the problem of Jammu and Kashmir should not been seen as sub-conventional or an insurgency; it is a proxy war by Pakistan. Accordingly, India needs to transform its strategy of deterrence so as to deter the Pakistan Army to continue its proxy war in Jammu and Kashmir.

Pakistan's strategy in Kashmir over the last 30 years leaves India with two options' first being to continue fighting the proxy war on our land, and second, to be proactive and shift from a strategy of 'deterrence by denial' to 'deterrence by punishment', and force Pakistan to review its strategy in Jammu in Kashmir. The second option has been discussed for quite some time within the Government, amongst the strategic community and within the defence forces. The argument of taking the war into Pakistan and resolving the Kashmir problem has been discussed as a ritual after 1947-48, 1965 and 1971 and was most vehemently taken up by the Armed Forces during the Kargil War 1999[65] and after every major terror strike. After crossing the nuclear threshold in 1998, Pakistan became more audacious in its proxy war, the magnitude and sophistication of the terror attacks increased, the Parliament attack in 2001, 2005, Delhi serial bomb blasts, 2006, Mumbai local train serial blasts, 2008, Mumbai attacks, 2010, Pune Bakery attack, Sunjuawn, Pakthankot, Kaluchak, Uri and Pulwama being some of them. Pakistan's newly acquired nuclear capability was one of the major reasons for escalation of the level violence in Jammu and Kashmir and also in commercial hubs like Mumbai, Pune and Delhi. The Pakistani military leadership understood the nuances of the nuclear deterrence and the apparent stability that comes when both adversaries are nuclear powers. Pakistan exploited the fact that India being the bigger and responsible power will have the onus to show restraint and avoid a total war which may escalate to a nuclear war. While India was stuck in the deterrence dilemma of how to react, aggressive diplomacy as a tool did not yield much results as Pakistan was also being considered a valuable player in the GWOT. It was only in 2014 that a surgical strike across the LoC on the terror camps in PoK indicated a shift in India's strategy of deterrence.

The apparent shift in India's strategy of 'deterrence by denial' to 'deterrence by punishment' had many scholars arguing that it may bring in some stability in South Asia by lowering the level of violence in Jammu and Kashmir. The cross-border strikes in the post-terror attack on the army camp in Uri in 2016 did not see much change on the ground; the infiltration and number of terror incidents kept going up.[66] However, a repeat of the 'deterrence by punishment strategy' in 2019, by carrying out aerial precision strikes on the terror camp at Balakot deep inside PoK, has seen a drop in violence and infiltration in the first half of 2019.[67] The world was forced to take notice, possibly for the first time the message was conveyed to the international audience and that in South Asia that India is not the only one to carry the baggage of being a restrained power or a rational nuclear state.[68] The impact was visible in the series of events that followed, the endorsement of the Indian action to strike the terrorist camp as a right to self-defence by many nations. The world took the note of the fact that the fundamental cause of instability in South Asia was Pakistan's policy of harbouring terrorist organisations and fighting a proxy war in Jammu and Kashmir. The international pressure mounted on Pakistan through the likelihood of FATF sanctions and UNSC sanctions on terrorist groups and Masood Azhar. Pakistan reportedly has closed a few terror camps in PoK and arrested 2008 mastermind Hafiz Saeed. But it is still too early to reach any definite conclusion. This shift in India's strategy to deterrence by punishment was also visible against China during the standoff at Doklam, where the resolve shown by the Indian Army can be attributed to force accretions that have taken place since 2010.[69] This shift in strategy needs to be supported with commensurate force structuring and capability building, if the strategy is not to remain symbolic or end up being referred to as an election-winning manoeuvre. There is a need to have a separate strategy of deterrence by punishment for a hybrid war scenario. Some of the measures which may be considered to create a deterrence capability in a hybrid war scenario have been discussed in subsequent paragraphs.

Capability Building for Hybrid Warfare Scenario

The Indian Army Land Warfare Doctrine 2018, describes the threat to our nation as a combination of conventional threat on the borders and proxy war in the hinterland. This acknowledgment of the changing nature of a threat to hybrid war is bound to stay. Hybrid war is complex and needs comprehensive national power to counter it. The military as a sole responder to the hybrid threat may not be the best option. In this part of the chapter the recommendations for military and structural changes required at the national level have been discussed. The capability building for credible a deterrence for a hybrid scenario will involve conventional and unconventional domains as well as kinetic and non-kinetic domains. The recommendations being made are for the collusive hybrid threat from India and China.

Conventional Domain

- **Need for Joint Doctrine on Hybrid Warfare.** The Indian Army Land Warfare Doctrine 2018 and also the Joint Armed Forces Doctrine 2017, acknowledge the threat of hybrid war and collusion of China and Pakistan. However, both the doctrines do not specify any measures to be undertaken to prepare or as a response to the hybrid threat. While the primary task of the armed forces would remain the conventional external threat, with the blurring of the line differentiating external and internal threats, there is a need to train and organise the forces accordingly.
- **Achieve Cooperation.** The lack of cooperation between the civil and military establishments and between the services itself has been highlighted by a number of reports and scholars. The most notable was the Kargil Review Committee report and the Group of Ministers (GoM) Report. Some of the recommendations made by the GoM have been implemented. However, a number of important recommendations are yet to be implemented such as appointment of a Chief of Defence Staff (CDS). The lack of cooperation amongst the services is leading to individual prioritisation in force modernisation, planning and training thereby leading to duplication and inefficient resource utilisation. This becomes more apparent in the field as there are 'nineteen commands between the three services and the Integrated Headquarters but not one of them is co-located with the other; and there is no co-relation, linkage, etc.'[70] India's weakness on cooperation calls into question its ability to formulate a coherent tri-services response to the 'two-front threat' posed by ongoing territorial disputes with Pakistan and China.[71] There is an urgent need for reforms in the civil-military and within military services to achieve more integration.
- **Intelligence.** Intelligence is the first wall of defence in hybrid warfare. The cognitive domain is targeted before the physical domain. The vulnerabilities caused due to increasing stress created by religion, societal norms, culture, social media, increasing disparity and radicalisation need to be identified well in time. The mapping of human terrain requires intellectuals, social

scientists, cyber experts, information warriors and professionals to understand and unravel any uneven behaviour by individuals and groups of people.[72] Intelligence agencies need to function in an integrated manner to identify any such trigger well in time for the government to respond.

- **Special Forces.** Response to acts of hybrid warfare need to be surgical and result in high impact. The Indian Army's response to the terrorist attack on the army post at Uri by carrying out a surgical strike on terrorist camps across the Line of Control in PoK and later air strike at a terrorist camp indicated a change in the Indian strategy to 'zero tolerance.' Well-trained, equipped Special Forces with the capability to insert by air is mandatory to add credibility to the 'zero tolerance' strategy.
- **Long-Range Precision Strike Capability.** The long-range precision strike capability can also achieve the desired impact as Special Forces, albeit with little loss. Such weapons also reduce decision and response time to achieve the desired impact. There is a need to invest more in long-range precision strike capability.
- **Anti-Satellite Strike Capability (ASAT).** The information domain plays one of the most important links in the execution of hybrid war. The capability to destroy the communication satellite system in space will boost the deterrence capability of the state. India fired the technology demonstrator 'Mission Shakti' on March 27, 2019 by conducting an anti-satellite missile test from the Dr. A P.J. Abdul Kalam Island launch complex. As this was a technology demonstrator, there is now a need to graduate to its space deterrence strategy along with its associated requirement—space doctrine. India needs to build capability to credibly deter Beijing's space denial programs below the level of ultimate physical violence directed at various space systems—the gray zone in which more counter-space activities are likely to materialize in the future.[73]
- **Cyber Security Doctrine.** India promulgated a National Cyber Security Policy in 2013 with an integrated vision and strategies for implementation. In practice, the application of this policy remains dispersed (fragmented) across different sectors.[74] A National Cyber Security Coordinator (NCSC) under the National Security Council Secretariat (NSCS) has been established who coordinates with different agencies at the national level on cyber security matters. Then there is a National Cyber Coordination Centre (NCCC) which is under the MHA, which has been set up to generate necessary situational awareness of existing and potential cyber security threats and enable timely information-sharing for proactive, preventive and protective actions by

individual entities. However, these agencies do not have the power to enforce or take action against the violators of cyber security. There is also a lack of synergy and convergence. There is a need to have a cyber-security doctrine, which will give a direction to the cyber policy and bring different agencies under one umbrella.

Unconventional Domain

Create Capability to Carry out Subversion in Pakistan. As India and Pakistan historically shared the same culture, tradition, language and religions, the internal fault lines in Pakistan are almost similar as in India. These fault lines need to be exploited or managed so that they can be leveraged when required. There are at least four major insurgency/secessionist/autonomy movements in Pakistan amongst other fault lines like sectarian divisions and minority dissent. These movements require an international audience, media coverage, social media outreach and publicity which does not require much effort these days in networked societies. At a higher level, leadership support, guidance, financial, political and other support can also be provided. Basic information about these movements are given below:

- **Baluchistan Movement.** Baluchistan is the largest province accounting for 42 percent of the land mass. However, it has the least population density. The paradox of a Baluchistan is that in spite of being the richest province in terms of natural resources, economically it remains one of the most backward. The province has been demanding independence, has fought a number of violent struggles from 1948 (first one after independence), and in 1958, 1962-68, and 1973 to 1975 and till date. In 2009, the Khan of Kalat (leader of the independence movement) declared himself as the leader of Baluchistan and announced a council of ministers for independent Baluchistan.[75] The province has been exploited since the formation of Pakistan. Even now, when Gwadar port and SEZs are being created, the benefit of the CPEC is being denied to the province. Against the exploitation and marginalisation, there have been a large number of protests and violent representations. Unrest in Baluchistan will hurt both Pakistan and China due to rising stakes in the CPEC and Gwadar port.
- **Internal Fissures in Punjab.** Punjab is the second largest, most densely populated and most powerful state in Pakistan. Punjab may look as the state from which the Pakistan nation draws its strength. However, there

are fissures like separatist movement of Sirkari-speaking people, dissent between North and South Punjabis and the increasing dominance of hardliners radical/terrorist groups. Sirkari is a distinct language spoken by approximately 10 percent of Pakistan's total population in both Punjab and Sindh. They are protesting with a demand for an independent state. Within Punjabis, there is animosity between the North and South Punjabis. One of the emerging fault lines in Punjab is the rise of radicalization and surge of radical terrorist organisations. The headquarters of the Tabligh-i-Jamaat in Raiwind and Lashkar-e-Tayyeba (LeT) in Muridke are located in Punjab, apart from the mushrooming of madrassas in the region. Recruitment for terror outfits has become an industry in Punjab. This has created a catch-22 situation in Pakistan. Any anti-terrorist campaign now creates ripples and turbulence in Punjab.

- **Federally Administered Tribal Area (FATA).** FATA remains one of the least backward areas of Pakistan with a literacy rate below 20 percent and availability of sanitation to even less people. This area has been traditionally under indirect rule, both by the British and later under Pakistan due to the fiercely independent character of the ethnic tribes. The dynamics of the region changed with the Russian invasion of Afghanistan, and subsequent adaptation and encouragement of Islamic terrorism as the preferred option to fight the Russians in Afghanistan. After 2001 and during GWOT the region once being considered as a pivot against the Russian occupation of Afghanistan ended up being the battleground and target of drone strikes. The collateral damage due to the infamous Pakistani operation of Zarb-e-Azb and the US offensive against Al Qaeda has left a deep sense of anguish and victimization amongst the local populace.[76]
- **Pashtun Tahaffuz Movement (PTM).** The Pashtuns are the second largest ethnic group in Pakistan and are concentrated in the north-western part of the country, sharing the Afghanistan border. The interesting fact is that the Pashtuns are the largest community in Afghanistan also. The Pashtun Tahaffuz (Protection) Movement (or PTM), was started by human rights activist Manzoor Pashteen to primarily raise the issue of human rights violations being carried out by the Pakistani Army. The Pashtun community has been the biggest casualty of the anti-terrorist crackdown by the Pakistan Army. The movement is just about a little more than a year old, but has generated a lot of traction. It has a

tremendous potential and can be a game-changer in Pakistan's internal security situation due to the following reasons:

- The Pashtun ethnic group has one of the largest representations in the Pakistan Army, accounting for between 15 and 22 percent among officers, and between 20 and 25 percent among the rank-and-file.
- Till date, eight COAS of the Pakistan Army have been Pashtun, Generals Ayub Khan, Yahya Khan, Gul Hasan Khan and Waheed Kakar.
- The idea of Pashtunistan is a movement which is even older than the existence of Pakistan. It spans across the border into Afghanistan. The Durand line was never accepted by the Pashtuns as an international border. Pakistan was able to suppress the movement but the PTM movement has the potential to revive the Pashtunistan movement.

- **Sindhi Dissent.** Sindh is the second-most populous province of Pakistan. Sindhis have a strong socio-cultural identity, rich literature and a distinct script. The Sindhi community is facing the threat of losing its identity amongst the Mohajirs and Punjabis. The feeling of alienation and marginalisation has led to insurrection, which surfaced in 1983 when popular Sindhi leader Zulfiqar Ali Bhutto was assassinated by Zia-ul-Haq.[77] The movement started for the creation of 'Sindhudesh', however, due to lack of international support, has died down. The Sindhi nationalists also created a Sindhi National Army, which undertook a number of strikes against the government. Baluchi and Sindhi activists have undertaken strikes jointly against the government.[78] As of today, the movement lacks external support and international attention. It also lacks an organizational set-up and leadership. This is an opportunity in waiting; the potential of this movement lies in creating instability in Karachi, which is the capital of Sindh province and the financial capital of Pakistan.
- **Mohajir Autonomous Region.** The term 'Mohajir' loosely denotes those Muslims who decided to migrate to Pakistan from India during independence. These people are largely settled in and around Karachi. The Mohajirs soon dominated Karachi and that led to marginalisation of the ethnic Sindhi community in their own province. The movement for an autonomous 'Greater Karachi' province has been simmering since the 80s. The political party, 'Mohajir Qaumi Movement' (MQM)

representing this community also has a powerful military wing which has been responsible for large-scale ethnic violence.[79]

To Exploit the Fault Lines in China. 2019 has not been a very good year for China with the trade war with the USA leading to Chinese economic growth rate recording the lowest figures in the last 30 years.[80] The worst is yet to come with increasing external instabilities and uncertainties and the internal unbalanced development and with many international companies leaving the Chinese economy leading to its downward pressure.[81] The pro-democracy demonstration in Hong Kong in July 2019 has brought out the suppressed feelings of the citizens against Communist rule. The ripples of the protests in Hong Kong are likely to influence the Taiwan nationalists too. Taiwan is assisting and providing shelter to the protesters of Hong Kong seeking asylum.[82] This is the suitable time to support the democratic movements in Asia, being the largest democracy in the world. The known internal fault lines in China, i.e., Tibet and Xinxiang issues are being controlled with an iron hand by China. However, these movements are like dormant volcanoes, waiting to explode. In Xinxiang, the cultural detention camps being run by the PLA have generated a lot of resentment in the world Muslim community. Both the movements need leadership and organizational support, which provides an opportunity to build leverage against China.

Creation of Offensive Cyber Warfare Capabilities. Cyber war is a new weapon of mass destruction. The characteristics like anonymity, lack of attribution, mass effect and minimal financial outlay, make cyber warfare one of the most attractive weapons of hybrid warfare. Whether it was a stuxnet attack on Iran's nuclear program in 2010 or the recent cyber-attack by the USA on Iran in retaliation to the shooting down of a drone, it indicates the preference of world powers to use cyber as an offensive weapon. Indian expertise in the cyber domain is well known, India needs to convert and use its strength in the cyber domain in such a manner that it becomes one of powerful tools of deterrence.

Organisational and Structural Reforms. Building of deterrence against hybrid warfare will require certain organizational and structural reforms. The hybrid threat does not threaten the borders only; rather it targets the arteries, brain, senses and the response mechanism of the nation simultaneously. A complete government approach is the only way to be prepared to tackle hybrid threats. Singapore follows a model of total defence, which has been very successful in defending this multi-cultural society. As Singapore is a city state, hence its cohesiveness is of a very high order. While the environments of Singapore and India can never be compared, still the study of their model brings out a number of valuable outcomes which will help in formulating a framework for other nations.

> *"The price of freedom is constant vigilance against threats new and old...... Whether it is against terrorism, cyber-attacks or border intrusions, we need total defence to remain strong and sovereign. Security threats can be real and physical like terrorism or just as damaging can come through the cyber world. Malicious malware can cripple our systems, fake news can cause racial riots and divide our people. Through total defence we have prepared ourselves well against threats in physical world over the past 35 years. This year to respond to threats in the digital domain, we will add the sixth new pillar to our total defence frame work, the digital defence."*
>
> **Dr. Ng Eng Hen, Minister for Defence, Singapore**
> 15 February 2019 (National Defence Day)

Singapore Model of Total Defence

Singapore's comprehensive security strategy known as 'Total Defence' was launched in 1984. The concept involves every Singaporean playing a part, individually and collectively, to build a strong, secure and cohesive nation. '*When we are strong, we are able to deal with any crises.*[83] The concept based on the empiricism that the threat to the nation does not manifest on the border only, the enemy can also target the economic systems, digital networks, or destroy the social fabric through non-traditional combat means. Total defence caters for a response to all-round threats and challenges. It has six pillars, military, civil defence, economic, social, psychological and digital. The sixth pillar, digital was added in 2019. The program is dynamic and evolves with the changing threat scenario. The individual pillars are discussed in brief below:

- **Military Defence.** This pillar is about keeping Singapore secure against any external threat. The Singapore Armed Forces (SAF) depends on the commitment of its National Servicemen along with the support of their families and employer.[84]
- **Civil Defence.** The Ministry of Defence, Singapore, describes civil defence as *"The ability to spot signs of threats, respond effectively and recover quickly from crises."*[85] *Civil Defence emphasizes that society is always alert, they know what to do in an emergency, natural calamity or a terror strike. The Singapore Civil Defence Force (SCDF) conducts emergency preparedness programmes and emergency exercises.*

- **Economic Defence.** The Ministry of Defence, Singapore, describes economic defence as "*A strong and resilient economy that is globally competitive and able to bounce back from any crises.*"[86] *Economic defence is emphasized as keeping the economy strong, resilient and have the ability to recover quickly whenever confronted by a challenge or crisis. The aim is to retain the confidence of investors in Singapore.*
- **Social Defence.** Singapore is a multiracial, multi-religious and multi-cultural society. Peace and progress is only possible in such a society when Singaporeans of all races and religions live together in harmony and look after one another. Social defence is visualized as everyone making an effort to trust one another and to strengthen the bonds across the different ethnic groups so that Singaporeans are strong and united especially during times of national challenges.[87]
- **Digital Defence.** The Ministry of Defence of Singapore describes strong digital defence as '*every individual is the first line of defence against threats from the digital domain. There is a need to build robust defence and have effective recovery plans to remain resilient even when things go wrong.*'[88] It emphasizes the role of each individual in digital defence, and towards empowering them there are programs and schedules to train individuals in cyber security and computer hygiene.
- **Psychological Defence.** Psychological defence is described as the ability of every individual and society as a whole to overcome threats and challenges, and to have the resolve to stand up for Singapore when pressured by forces that undermine our national interests, and the fighting spirit to press on and overcome crises together.[89]

The program is run under the National Education & Community Engagement department under the Ministry of Defence of the Singapore Government. The department acts as the nodal agency to run various programs, training schedules and information campaigns to educate people about their role and responsibility towards the security of the nation. A glance at the web site of 'Total Defence' shows the various activities conducted in 2019 by the department.

- These activities range from short programs on psychological defence course, first aid, emergency response, cyber security and to teaching of new skills to become economically independent.
- It also manages the 'Harmony Fund' which is used for running programs which increase harmony between various religions.

- There is another fund under this program called 'Singapore Fund', which supports initiatives which build national identity or serve the need of the community by individuals or NGOs.
- It manages the Singapore Armed Forces Volunteer Corps, which is a volunteer scheme to provide more opportunity to civilian citizens or permanent residents to contribute to national defence, show support for National Service, and deepen their understanding and ownership of national defence.

Takeaways from Singapore Model of Total Defence. Singapore is a city state, the challenges are different when compared to a large country like India. But at the same time the redundancy or capacity to bear a shock is also less accordingly. There are a few similarities between India and Singapore: both nations are multicultural, multi-racial and multi religion societies. There are a number of takeaways from the Singapore model, which can be suitably modified and adapted for consideration here.

- First and foremost is acknowledgment of the nature of the threat.
- Realisation and understanding of the fact that emerging security threats can only be fought successfully when every citizen is ready to play his part.
- Identification of various pillars which form part of comprehensive security. This may vary from country to country.
- Role of Government in educating and partnering with the populace to face impending threat and security challenges.
- A nodal ministry or statutory body to monitor and coordinate the activities between concerned ministries. In Singapore, the Ministry of Defence is the nodal agency.
- Allocation and creation of funds to manage internal fault lines. Singapore has identified the multi-cultural/multi-religious community (four communities—Chinese, Indian, Muslim, and Malay) as potential fault lines. Accordingly, the Harmony Fund has been created which focuses only on the conduct of harmonious activities between these communities.
- Regular review and reform of the 'Total Defence Concept', like considering an increase in cyber espionage and cyber-attacks. Digital defence was added as the sixth pillar in 2019.

'There is both danger and clinging to any long-established status quo.... Procrastination has cost nations dear'

—Kargil Review Committee Report, 1999

Recommendations for Organisational and Structural Reforms

India's national security apparatus has evolved over the last 70 years. The various setbacks that the country has suffered in the last 70 years necessitated the creation of different institutions, which have matured over a period of time. The sequential creation of such institutions has also led to the duplication or overlapping of their respective charters of responsibilities. As the Kargil Review Committee, noted the status quo in security arrangements would lead to a heavy cost in the long run. Hence, it is imperative that the security apparatus of the nation is refined and adapted to the changing threat scenario. Some recommendations with respect to organisational and structural reforms are given below:

- **Comprehensive National Security Strategy.** There is a need to formulate a National Security strategy which will also bring out the vertical and horizontal escalation matrix in case of any hybrid attack against India. This is imperative to have a credible deterrence strategy against hybrid warfare.
- **Creation of Nodal Agency.** As discussed in earlier chapters, internal fault lines are exploited by adversaries in the domain of hybrid warfare. Hence, a holistic government approach is the only way to reduce the vulnerability against hybrid war. In order to achieve convergence and a common understanding of national security and to coordinate the efforts of various departments of the government, there is a need to have a nodal agency created by an Act of the Constitution. The Cabinet Committee on Security (CCS) or National Security Council (NSC) or the enhanced Strategic Policy Group (SPG) or a separate body (newly created) could be made the nodal agency to coordinate the entire government approach. The recommended nodal agency should be created by an Act of Parliament and have the power to implement decisions through the concerned ministry. The NSC in the current form is an advisory body, with no powers to implement decisions on national security.[90] The cabinet ministers of External Affairs, Home, Defence and Finance should be members of this agency so as to ensure implementation of the directions of the nodal agency.

- **Core Group of Important Ministries.** The Cabinet Committee on Security clears all major decisions related to the security of the nation. However, the committee meets as on an as required basis depending on the exigencies of any given situation. Having analysed the definition, nature, scope and existence of hybrid threat on the well-being of the nation, there is a need to have a core group of important ministries which will be responsible to monitor the security situation on a regular basis and carry out periodical reviews. Considering the dynamic and ever evolving nature of threats, an episodic model of security establishment at the highest level may not be the best option. Currently, the CCS has representation of four ministries like Defence, Home, Finance and External Affairs. However, the nature of threat today covers a much wider spectrum. Hence, there is a need to involve the ministries of Human Resource Development, Health and Information & Broadcasting also. A pictorial representation of the nodal agency and core group of ministries is given below:

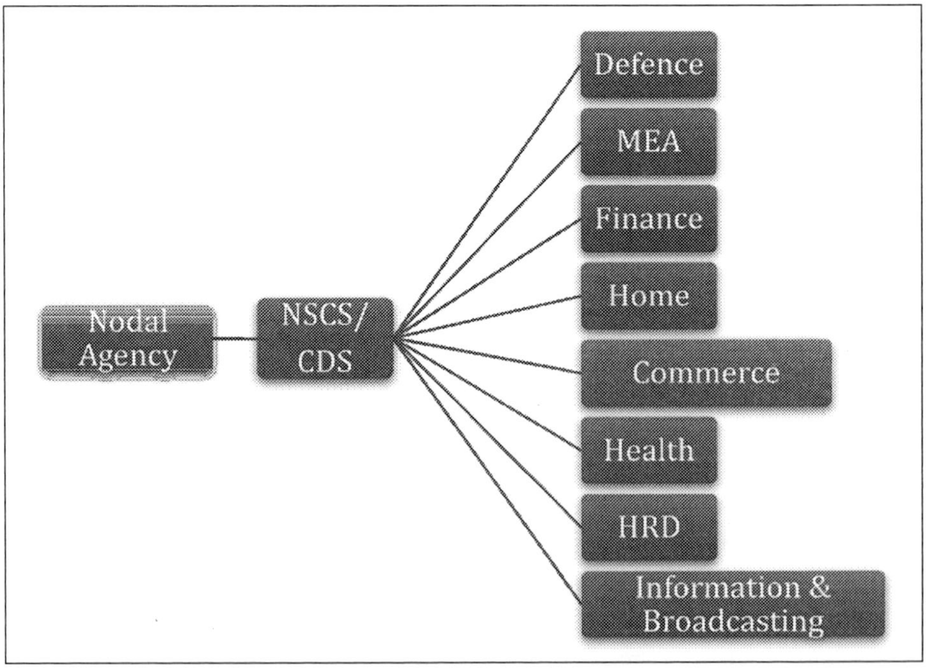

- **Role of CDS.** The long-pending necessity and one of the key recommendations of the Kargil Review Committee, GOM report and the Naresh Chandra Committee reports regarding the appointment of a CDS was announced by PM Shri Narendra Modi on Independence Day. Gen Bipin Rawat took over as the first CDS of country in Jan 2020. The operational responsibilities of CDS were promulgated vide Ministry of Defence circular of 20 January 2020.[91] The Department of Military Affairs has been created which will be headed by CDS. One of the models could be that the office of the CDS be incorporated with the nodal agency and be made the lead to take carry out a yearly review of the security situation or the threat perception. This yearly review or threat perception will then become the guiding document for the various key ministries. The nodal agency may also become the lead to prepare the Draft National Security Strategy.

- **Empowered National Cyber Security Agency.** Considering the emerging importance and threat from the cyber domain, there is a need to have an umbrella organisation under which all the agencies concerned with cyber security should function. The current National Cyber Security Coordinator can be notified and empowered by an Act of the Constitution.
- **Integration of Cyber, Space and Electromagnetic Warfare Assets.** Cyber, space and electronic warfare, by their nature and commonality of assets and resources, cannot be separated from each other. The Naresh Chandra Task Force in 2012 had recommended the integration of all such assets under one umbrella to execute operations in tune with the strategic objectives of national interests.[92] The PLA, has achieved the necessary synergy and coordination in its information warfare by creating the Strategic Support Force. This organisation has been formed by integrating all the units which were dealing with cyber, space and electronic warfare. The creation of the Cyber Agency, Space Agency and Special Forces Division in 2018-19 is the first step towards integration. However, the forces still remain in separate independent verticals. There is a need to create another organisation like the Strategic Forces Command, which will encompass cyber, information and space warfare assets.
- **National Counter Terrorism Centre (NCTC).** The hypothesis behind the setting up of the NCTC as a single, unified counter-terrorism agency was to synergise and optimise national efforts and resources for safeguarding the nation against terrorism. For more than 10 years now,

the proposal is stuck in political perplexity. In 2018, the Parliamentary Standing Committee on Home Affairs, in a report to the Rajya Sabha, recommended that the Government should immediately revive the idea of a National Counter Terrorism Centre (NCTC) as a unified counter terrorism agency.[93] The report also said that while there were concerns of some states about the provisions, the MHA should take the lead to address the concerns and build up a consensus with regard to the structure, functions, mandate and utility of creating and operationalising of the NCTC as a single, unified and premier counter-terrorism agency.

- **Operationalisation of NATGRID.** Intelligence is the first line of defence in combating terrorism and to secure the nation from any kind of internal and external threat. In the current environment, technology has transformed the way Intelligence is collected, collated, processed and analysed. The speed, medium and sources like satellite, electronic, social media (open source Intelligence) require a high degree of coordination and integration. Setting up and operationalisation of NATGRID is an essential step towards that aim.
- **Police Reforms.** The police force is the cornerstone of strategy to combat terrorism and other internal security threats. In India, it is primarily trained to maintain law and order and not so well trained to tackle terrorism. While the police has probably the best grassroots Intelligence, it does not have resources or manpower to act. It is also not so well networked with the Central intelligence agencies and forces which have the strength, resources and training to deal with terrorism and other emerging challenges. Police reforms are possibly one of most debated and long-pending reforms. There is an urgent need to undertake holistic reforms to include organisational structure, equipment, training and budgeting. As hybrid war is emerging as the biggest threat to the nation, police reforms becomes one of the key focus areas for a functional strategy to combat terrorism.

NOTES

1. Indian Army Land Warfare Doctrine-2018, available at http://www.ssri-j.com/MediaReport/Document/IndianArmyLandWarfareDoctrine2018.pdf, accessed on July 25, 2019
2. For a detailed description of land borders, see site of Border Management Division of Ministry of Home Affairs of GOI, can be accessed at https://mha.gov.in/division_of_mha/border-management-i-division#skipCont

   Also see chapter on Border Management in book by Dr. Arvind Gupta, '*How India Manages its National Security'* Penguin Random House, 'p. 184

3. Dr Arvind Gupta, *How India Manages its National Security*, Penguin Random House, p. 194
4. Annual Report for 2017-18, Ministry of Home Affairs, available at https://mha.gov.in/sites/default/files/MINISTRY%20OF%20HOME%20AFFAIR%20AR%202017-18%20FOR%20WEB.pdf accessed on 29 June 2019
5. Ibid
6. Pushpita Das, in *Coastal Security Indian Experience*, monograph, IDSA, available at https://idsa.in/system/files/Monograph22.pdf, accessed on 15 June 2019
7. Ibid
8. Dr Arvind Gupta, *How India Manages its National Security*, Penguin Random House, p. 19
9. Please refer to Chapter 3, where in the rise of Islamic terror groups and the pivotal role played by Pakistan has been analysed in detail.
10. "18 Years of Fighting, 2400+ American Lives, A Trillion Dollars And The Taliban Is Back", web article in Strategic News International (SNI), available at http://sniwire.com/neighbours/18-years-of-fighting-2400-american-lives-a-trillion-dollars-and-the-taliban-is-back/ accessed on Jul 14, 2019
11. Shubhankar Basu, in Revival of Peace Initiatives in Afghanistan: Implications for India, Issue brief, VIF, available at https://www.vifindia.org/brief/2019/january/31/revival-of-peace-initiatives-in-afghanistan-implications-for-india accessed on 05 Jul 2019
12. For a detailed analysis of the change in demographic profile of bordering states, see Dr. Pushpita Das, *Illegal Migration from Bangladesh*, monograph, IDSA, available at https://idsa.in/monograph/illegal-migration-from-bangladesh_pdas, accessed on 3 July 2019
13. Bangladesh Population Review, *World Population Review*, available at http://worldpopulationreview.com/countries/bangladesh-population/, accessed on 2 July 2019
14.. China's Growing Influence In Nepal vs India's Soft Power, *Outlook* India, web edition, Feb. 9, 2019, available at https://www.outlookindia.com/blog/story/india-news-chinas-growing-influence-in-nepal-vs-indias-soft-power/4039 accessed on 3 July 2019
15. Nepal Inching Towards China, in *Diplomat*, 9 May 2019, available at https://thediplomat.com/2019/05/nepal-inching-toward-china/ accessed on 2 July 2019
16. Nepal backs out of BIMSTEC joint military exercise in Pune, *Indian Express* Web Edition, Sep. 9, 2018, available at https://indianexpress.com/article/india/nepal-backs-out-of-bimstec-joint-military-exercise-in-pune-5346815/, accessed on 1 July 2019
17. Pakistan's ISI gets new control room in Nepal to carry out terror activities in Kashmir, in Zee News. Com, available at https://zeenews.india.com/india/pakistan-s-isi-gets-new-control-room-in-nepal-to-carry-out-terror-activities-in-kashmir-2212358.html, accessed on July 2, 2019
18. Corridor of Terror, *Outlook*, web edition, Jan. 17, 2000, available at https://www.outlookindia.com/magazine/story/corridor-of-terror/208719, accessed on 3 July 2019
19. Ibid
20. Lt. Gen. Prakash Katoch, Nepal – fast becoming ISI Control Centre? *Indian Defence Review*, web edition, May 11, 2018, available at http://www.indiandefencereview.com/spotlights/nepal-fast-becoming-isi-control-centre/, accessed on 2 July 2019
21. Dr. Arvind Gupta, *How India Manages its National Security*, Penguin Random House, p. 27
22. Ibid
23. For Sri Lanka, a Long History of Violence, *The New York Times*, web edition, April 21, 2019, available at https://www.nytimes.com/2019/04/21/world/asia/sri-lanka-history-civil-war.html,

accessed on 2 July 2019

24. For Sri Lanka, a Long History of Violence, *The New York Times*, web edition, April 21, 2019, available at https://www.nytimes.com/2019/04/21/world/asia/sri-lanka-history-civil-war.html, accessed on 2 July 2019
25. For more details on the history of Jammu and Kashmir, see Jagmohan Singh, *My Frozen Turbulence in Kashmir*, Allied Publishers, India
26. Naxalism biggest threat to internal security: Manmohan, *The Hindu*, May 10, 2010 available at https://www.thehindu.com/news/national/Naxalism-biggest-threat-to-internal-security-Manmohan/article16302952.ece accessed on Dec 10, 2018
27. Refer to PQ Number, LS.US.Q No 489, dated 25.06.2019
28. Stephen P. Cohen and Sunil Dasgupta, Arming without Aiming: India's Military Modernization (Washington, DC: Brookings Institution Press, 2010), pp. 151–55.
29. Arvind Gupta, *How India Manages its National Security*, Penguin Random House, p. 216
30. Ibid
31. Ministry of Home Affairs, Annual Report 2017–18, https://mha.gov.in/sites/ default/files/ MINISTRY%20OF%20HOME%20AFFAIR%20AR%202017-18% 20FOR%20WEB.pdf
32. Sarah J. Watson and C. Christine Fair, "India's Stalled Internal Security Reforms," *India Review, 12*, No. 4 (2013): 280–99
33. National Intelligence Grid to be ready by early 2020, *India Today*, Sep. 22, 2019
34. Ibid
35. For a detailed report, see ' Income inequality gets worse; India's top 1% bag 73% of the country's wealth, says Oxfam,' *Business Today*, web edition, Jan. 30, 2019, available at https://www.businesstoday.in/current/economy-politics/oxfam-india-wealth-report-income-inequality-richests-poor/story/268541.html, accessed on July 5, 2019
36. For details, see 'Reporter asks DMK's MK Stalin about 'Dravida Nadu' movement. Here's his reply,' *India Today*, web edition, Mar. 17, 2018, available at https://www.indiatoday.in/india/story/reporter-asks-dmk-s-mk-stalin-about-dravida-nadu-movement-here-s-his-reply-1191903-2018-03-17, accessed on July 5, 2019
37. 'For more details on the report, see' 58 terror modules operating from illegal madrasas in West Bengal, reveals probe', *India Today*, web edition, Oct. 31, 2014, available at https://www.indiatoday.in/india/east/story/west-bengal-madrasas-tterror-modules-burdwan-blast-bangladesh-nia-nsg-ib-225251-2014-10-31, accessed on 28 Dec. 2018
38. The Steering Committee. "Making India 5G ready" 5G High Level Forum. August 23, 2018 available at : http://dot.gov.in/sites/default/files/5G%20Steering%20Committee%20 report%20v%2026.pdf?download=1, accessed on 4 July 2019
39. Ibid
40. 'US launched cyber-attack on Iranian rockets and missiles – reports,' in *The Guardian*, web edition, June 22, 2019 available at https://www.theguardian.com/world/2019/jun/23/us-launched-cyber-attack-on-iranian-rockets-and-missiles-reports, accessed on 4 July 2019
41. Terrorist use of crypto currencies, Rand Publication, https://www.rand.org/content/dam/rand/pubs/research_reports/RR3000/RR3026/RAND_RR3026.pdf
42. Ibid
43. General V.P. Malik, PVSM, AVSM (Retd), in the 30th USI National Security Lecture A Comprehensive Response Strategy to a Collusive and Collaborative Threat from China and Pakistan, 2014

44. As narrated by Lt. Gen. (Retd) Ravi Sawhney, former General Officer Commanding, 4 Corps, during an interview with the author on July 11, 2019
45. This quote has been taken from an article written by former Ambassador to Pakistan, Shri G. Parthasarthy. For the article, see 'Beware the China-Pakistan nuclear axis,' *Business Lline, The Hindu*, web edition, Nov. 1, 2017, available at https://www.thehindubusinessline.com/opinion/beware-the-china-pakistan-nuclear-axis/article22220540.ece1, accessed on 5 July 2019
46. Beware the China-Pakistan nuclear axis,' *Business Lline, The Hindu*, web edition, Nov. 1, 2017, available at https://www.thehindubusinessline.com/opinion/beware-the-china-pakistan-nuclear-axis/article22220540.ece1, accessed on 5 July 2019
47. Ibid
48. For more details on the China-Pakistan Nuclear Cooperation, see T. V. Paul in 'China-Pakistan Nuclear/Missile Ties and the Balance of Power,' *The Nonproliferation Review*, Summer 2003 available at https://www.nonproliferation.org/wp-content/uploads/npr/102paul.pdf, accessed on July 6, 2019
49. Ibid
50. Beware the China-Pakistan nuclear axis,' *Business Line the Hindu*, web edition, Nov. 1, 2017, available at https://www.thehindubusinessline.com/opinion/beware-the-china-pakistan-nuclear-axis/article22220540.ece1. accessed on 5 July 2019
51. General V.P. Malik, PVSM, AVSM (Retd), in the 30th USI National Security Lecture A Comprehensive Response Strategy to a Collusive and Collaborative Threat from China and Pakistan, 2014. The Secret Missile Deal, *Time*, June 30, 1997, available at http://content.time.com/time /magazine/article/0,9171,986606,00.htm, laccessed on July 4, 2019
52. Ibid
53. Pakistan ceded 5,180 km of Indian territory in Shaksgam valley in Pakistan Occupied Kashmir to China, under the so-called "Agreement between China and Pakistan on the boundary between China's Sinkiang and the contiguous areas, the defence of which is under the actual control of Pakistan" signed in March 1963. For more details, see Parliament Starred Question number 339 dated Dec. 14, 2006. Available at https://www.mea.gov.in/rajya-sabha.htm?dtl/8120/q+339+transkarakoram+pass, accessed on July 5, 2019
54. For more details of China's support to Pakistan, see Dr. Monica Chansoria, in 'China's Current Kashmir Policy: Steady Crystallisation', *CLAWS Publication*,Winter 2016
55. Ibid
56. 'Stop Blaming Pakistan and China: Chinese Media to India,' *Economic Times*, web edition, Feb. 18, 2019 available at https://economictimes.indiatimes.com/news/defence/stop-blaming-china-pakistan-chinese-media-to-india/articleshow/68050309.cms, accessed on July 6, 2019
57. Gilgit-Baltistan: Pakistan's Geopolitical Loophole, in The Diplomat, web edition, Feb 16, 2019, available at https://thediplomat.com/2019/02/gilgit-baltistan-pakistans-geopolitical-loophole/ accessed on Jul 10, 2019
58. Union Minister of state for Home Ministry Shri Nityanand Rai told Parliament that the security situation has improved in the state of Jammu and Kashmir, the infiltration in the state has come down by 43% in the first half of 2019 when compared with the corresponding figure of last year. For more see, Infiltration in J&K down after Balakot strike: Govt Times of India, web edition, Jul 09 2019, available at https://timesofindia.indiatimes.com/india/infiltration-in-jk-down-after-balakot-strike-govt/articleshow/70141971.cmsaccessed on Jul 10, 2019

59. Rajesh Rajagopalan, in "India's clueless deterrence "strategy" in ORF editorial, available at https:/ / www.orfonline.org/expert-speak/india-clueless-deterrence-strategy/,accessed on 6 July 2019
60. Janice Gross Stein, in his scholarly work '*Rational Deterrence against Irrational Adversaries*' has quoted Jhon Bolton who said, "When you have a regime [like Iran] that would be happier in the afterlife than in this life, this is not a regime that is subject to classic theories of deterrence. Retaliation, for them, which would obliterate their society, doesn't have the same negative connotations for their leadership,". This appeared in *The New York Times Magazine*, 4 Nov. 2007. The article by Stein can be accessed at 'Complex Deterrence' strategy in Global Age, edited by T. V. Paul,. P M. Morgan and J. J. Writz, University of Chicago Press, Chicago and London, 2009, pp. 58-85
61. Chapter 3 deals in detail with the strategic culture of the Pakistani Army and the wars it has fought with India
62. Janice Gross Stein, in 'Rational Deterrence against Irrational Adversaries', 'Complex Deterrence' strategy in Global Age,' edited by T.V. Paul, P.M. Morgan and J.J. Writz, University of Chicago Press, Chicago and London, 2009, pp. 58-85
63. 'Deterrence Trap' has been explained by Janice Gross Stein in the scholarly work 'Rational Deterrence against Irrational Adversaries.' The article by Stein can be accessed at 'Complex Deterrence' strategy in Global Age, edited by T.V. Paul, P.M. Morgan and J.J. Writz, University of Chicago Press, Chicago and London, 2009, pp. 58-85
64. Ibid
65. A panel discussion organized by Brookings India, '20 Years after Kargil: Military operations, perceptions, and decision-making', moderated by Prof. Anit Mukherjee. The panelists included Gen. V.P. Malik, Chief of Army Staff, Indian Army (1997-2000); Lt. Gen. Mohinder Puri, Major General, GOC, 8 Mountain Division (1999); Air Marshal Narayan Menon, Air Officer Commanding (AOC), Jammu and Kashmir; Shakti Sinha, Private Secretary to Prime Minister Atal Bihari Vajpayee (1996-1999) and Indrani Bagchi, Diplomatic Correspondent, *The Economic Times* (1999). All the panelists concurred that the Armed Forces asked for permission to cross the LoC. However, the decision was taken not to cross it. The proceedings of the panel discussions can be viewed at https://www.brookings.edu/events/20-years-after-kargil-military-operations-perceptions-and-decision-making/
66. Refer to data in Table 3.1, Chapter 3, p. 68
67. Net infiltration in J&K reduced by 43%: Union Minister Replies in Lok Sabha, in *The Times of India*, web edition, July 9, 2019 available at http://timesofindia.indiatimes.com/articleshow/70148745.cms?utm_source =contentofinterest&utm_medium= text&utm_campaign= cppst accessed on July 10, 2019
68. For more views on impact of Balakote, see Ashley J. Tellis, in 'A Smoldering Volcano: Pakistan and Terrorism after Balakot,' Carnegie, available at https://carnegieendowment.org/2019/03/14/smoldering-volcano-pakistan-and-terrorism-after-balakot-pub-78593, accessed on July 4, 2019
69. Yogesh Joshi and Anit Mukherjee, 'From Denial to Punishment: The Security Dilemma and Changes in India's Military Strategy towards China,' *Asian Security*, Nov. 2018
70. Arun Prakash, Keynote Address, 'Proceedings of USI Seminar on Higher Defence Organisation (New Delhi: United Service Institution of India, 2007)
71. Stephen P. Cohen and Sunil Dasgupta, *Arming without Aiming: India's Military Modernization,* Washington, DC: Brookings Institution Press, 2010, pp. 151–55
72. Narender Kumar, 'War beyond Rules: Hybrid War and India's Preparedness,' *CLAWS Journal*, Summer 2017

73. Ashley J Tellis, in 'India's ASAT Test: An Incomplete Success,' April 15, 2019, Carnegie Endowment for Peace, available at https://carnegieendowment.org/2019/04/15/india-s-asat-test-incomplete-success-pub-78884, accessed on July 29, 2019
74. Vivekananda International Foundation, Task Force Report on Credible Cyber Deterrence in Armed Forces of India 2018, p. 52, available at vifindia.org,, accessed on July 29, 2019
75. Manekhaw Paper, by Col. H.P.S. Hansi, 'Faultlines in Pakistan and Implications for India,' *CLAWS*, available at https://www.claws.in/images/publication_pdf/1090930096_MP66 Hansi.pdf, accessed on June 25, 2019
76. FATA: Terrorists or victims of a covert war?, *Al Zajeera*, web edition, Sep. 14, 2016, available at https://www.aljazeera.com/indepth/opinion/2016/09/fata-terrorists-victims-covert-war-160912084500746.html
77. For details, see 'Pakistan's Other Trouble Spot,' *Indian Express*, web edition, Dec. 7, 2016, available at https://indianexpress.com/article/opinion/columns/pakistan-trouble-spots-partition-talibans-baloch-nationalism-sindh-nationalists-3069384/, accessed on July 6, 2019
78. Ibid
79. 'Expatriate Mohajirs demand creation of 'Greater Karachi' as an autonomous region within Pakistan', *The Hindu*, web edition, Jan. 22, 2019 available on https://www.thehindu.com/news/international/expatriate-mohajirs-demand-creation-of-greater-karachi-as-autonomous-region-within-pakistan/article26057505.ece, accessed on July 6, 2019
80. 'China's Economic Growth Hits 27-Year Low as Trade War Stings,' *The New York Times*, web edition, July 14, 2019, available at https://www.nytimes.com/2019/07/14/business/china-economy-growth-gdp-trade-war.html?login=email&auth=login-email, accessed on July 15, 2019
81. Ibid
82. 'Hong Kong protests inspire Taiwan to ramp up resistance to China,' in *Nikki Asian Review*, web edition, July 16, 2019 available at https://asia.nikkei.com/Opinion/Hong-Kong-protests-inspire-Taiwan-to-ramp-up-resistance-to-China, accessed on July 17, 2019
83. What is Total Defence? for more details, see website of Singapore Police Forces at https://www.police.gov.sg/community-programme/total-defence, accessed on July 10, 2019
84. To know more about Total Defence, visit Ministry of Defence Singapore website, https://www.mindef.gov.sg/web/portal/mindef/home, accessed on July 12, 2019
85. Ibid
86. Ibid
87. Ibid
88. Ibid
89. Ibid
90. Arvind Gupta, *How India Manages its National Security*, Penguin Random House, India, p. 349
91. The allocation of work and staff between the Department of Defence and newly created Department of Military Affairs was promulgated by vide MoD Circular/order dated 20 Jan 2020, Available at mod.gov/in/dod/oms_order accessed on 10 Feb 2020
92. 'Agencies take shape for Special Operations, Space and Cyber,' *The Times of India*, web edition, May 19, 2019, accessed on July 30, 2019
93. Report of Department-Related Parliamentary Standing Committee on Home Affairs 205th to Rajya Sabha, available at http://164.100.47.5/newcommittee/reports/EnglishCommittees/Committee %20on%20Home%20Affairs/205.pdf

Conclusion

> *"Unless India stands up to the world, no one will respect us;*
> *In this world, fear has no place; only strength respects strength."*
>
> **—Dr. A. P. J. Abdul Kalam**

This book examines the security challenges that India faces in a scenario where two of the traditional adversaries decide to collude. The increasing economic collaboration between China and Pakistan has taken the relationship to the higher level. The China-Pakistan nexus has been further deepened by the CPEC and will pose a great security challenge to India in the future. The project has already altered the dynamics in PoK. There is a possibility that Pakistan may end up being a vassal state of China. The other neighbours such as Nepal, Bangladesh, Myanmar and Sri Lanka are also endangering their strategic autonomy by seeking greater financial assistance from China. Chinese influence in the region is increasing every day. In the years ahead, India is likely to face more security challenges due to non-traditional and non-kinetic threats. Emerging vulnerabilities like cyber security, widening fault lines of inter-ministerial coordination, radicalisation, economic disparity, insecurity amongst minorities, vulnerabilities of sensitive data base, banking networks, railway and air transport systems are some of the issues which may trip any day or likely to be exploited by a determined enemy.

The question that needs to be asked is whether India is ready to face the challenge of a collusive threat in the hybrid domain? This book also examines how successful the conventional and nuclear deterrence have been. It has not been able to prevent Pakistan from undertaking four major wars and a prolonged proxy war which is being fought by the Indian Army for more than 30 years now. This book in Chapter Five argues for the need to build deterrence in the hybrid domain. Such a deterrence includes capacity building in offensive cyber operations and exploiting the fault lines of adversaries. The chapter also reiterates the long-pending need of developing a national security strategy.

To depend heavily on the armed forces is not the recommended strategy to combat hybrid warfare. There is an urgent need to follow the functional approach for institutional building as recommended in Chapter Five. The existing and emerging fault lines have to be managed by respective governmental organisations or ministries. Inter-ministerial differences and civil-military relations need to be mended on priority. Long-pending police reforms, higher defence military organisation, defence procurement and indigenisation of defence production are some of the areas which need immediate attention and resolution at the highest level.

The transformation in the character of war has been driven primarily by technology. Disruptive technologies such as artificial intelligence, quantum computing, 5G, swarming, digital financial instruments, gene editing and autonomous weapon systems have the potential to increase the lethality of irregular or non-state actors. These low-cost high-impact instruments may even influence the way responsible states opt to fight a war. There is a need to study the implications of these technologies on the future of conflict. The doctrines, organisations, equipment and training of forces need to be updated with the advancement of technology to be able to respond to the emerging threats effectively.

Index